Oxford **Mathematics**

Primary Years Programme

2

Contents

OXFORD
UNIVERSITY PRESS
AUSTRALIA & NEW ZEALAND

G000134578

This is a one.

This is 12 ones
OR 1 ten and 2 ones.

This is 120 ones
OR 1 hundred and 2 tens.

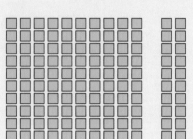

Guided practice

1 How many?

In a 3-digit number, the first digit is hundreds, the second is tens and the third is ones.

a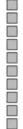

b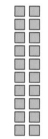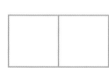

c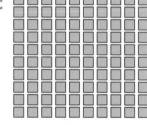

d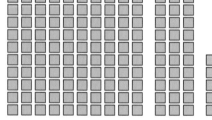

- -

2 Write the numbers.

a

b

c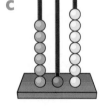

OXFORD UNIVERSITY PRESS

Independent practice

1 This is 354.

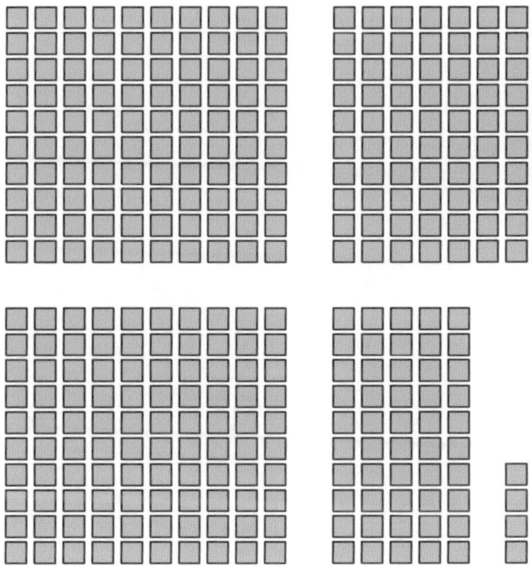

☐ hundreds

☐ tens

☐ ones

2 This is 206.

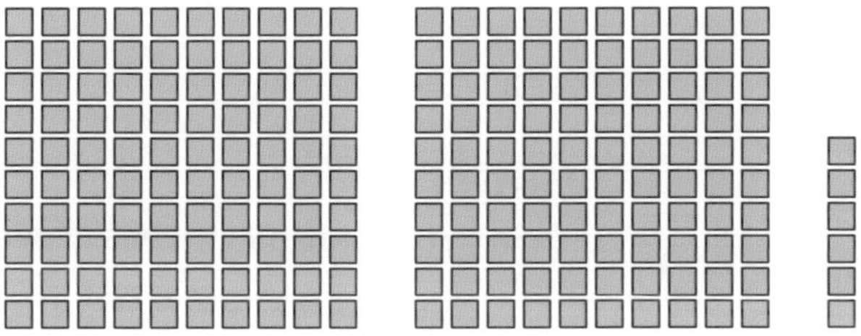

☐ hundreds

☐ tens

☐ ones

3 This is 423.

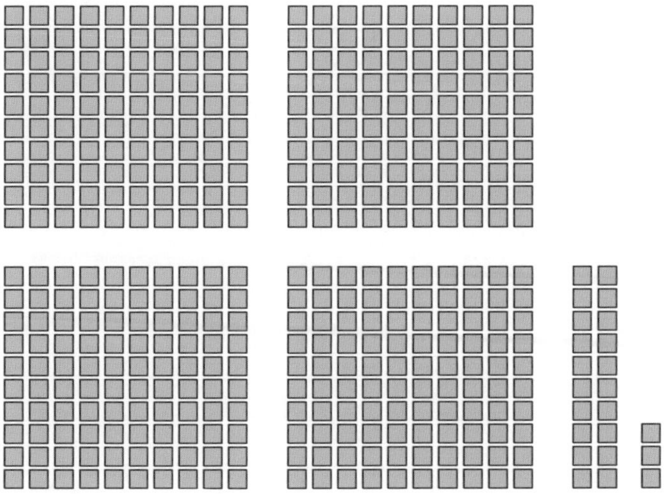

☐ hundreds

☐ tens

☐ ones

4

a In 849, the 9 is in the ones place. How many:

tens? ☐   hundreds? ☐

b In 347, the 4 is in the tens place. How many:

ones? ☐   hundreds? ☐

c In 413, how many:

hundreds? ☐   tens? ☐   ones? ☐

d In 508, how many:

tens? ☐   hundreds? ☐   ones? ☐

---

5 Draw 216.

How many different ways can you rename 216?

OXFORD UNIVERSITY PRESS

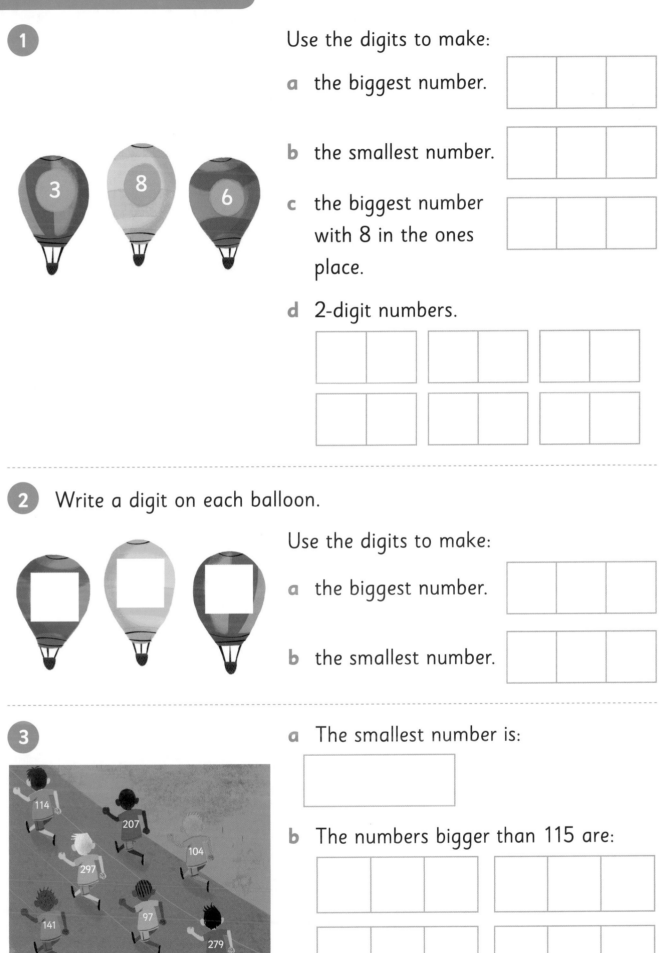

1

Use the digits to make:

a the biggest number.

b the smallest number.

c the biggest number with 8 in the ones place.

d 2-digit numbers.

2 Write a digit on each balloon.

Use the digits to make:

a the biggest number.

b the smallest number.

3

a The smallest number is:

b The numbers bigger than 115 are:

Adding with doubles

11 + 13 = 11 + 11 + 2 = 22 + 2 = 24

Guided practice

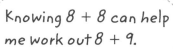

Knowing 8 + 8 can help
me work out 8 + 9.

1 Use doubles to help you add.

a   8 + 9 = 8 + 8 + ☐ = | 1 | 6 | + ☐ = ☐

b   10 + 13 = 10 + 10 + ☐ = ☐ + ☐ = ☐

c   15 + 17 = 15 + 15 + ☐ = ☐ + ☐ = ☐

d   12 + 13 = 12 + 12 + ☐ = ☐ + ☐ = ☐

e   14 + 15 = 14 + 14 + ☐ = ☐ + ☐ = ☐

OXFORD UNIVERSITY PRESS

1 Add these doubles.

a 7 + 7 = ☐☐   b 11 + 11 = ☐☐

c 16 + 16 = ☐☐   d 20 + 20 = ☐☐

e 25 + 25 = ☐☐   f 50 + 50 = ☐☐☐

2 Add these near doubles.

a 7 + 9 = ☐ + ☐ + ☐ = ☐☐ + ☐ = ☐☐

b 11 + 13 = ☐☐ + ☐☐ + ☐ = ☐☐ + ☐

= ☐☐

c 16 + 17 = ☐☐ + ☐☐ + ☐ = ☐☐ + ☐

= ☐☐

d 20 + 23 = ☐☐ + ☐☐ + ☐ = ☐☐ + ☐

= ☐☐

e 25 + 27 = ☐☐ + ☐☐ + ☐ = ☐☐ + ☐

= ☐☐

Getting to a 10

$28 + 6 = 28 + 2 + 4$

$\qquad = 30 + 4$

$\qquad = 34$

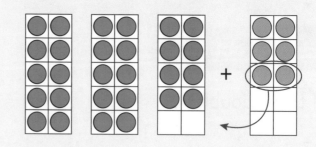

It's easier to add numbers to a tens number like 10, 20 or 30.

Guided practice

1 Add by getting to a 10.

a $17 + 6 = 17 + \boxed{3} + \boxed{}$

$\qquad = 20 + \boxed{}$

$\qquad = \boxed{}$

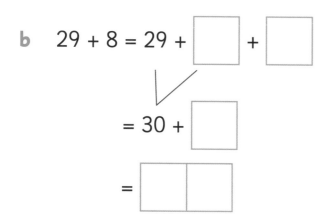

b $29 + 8 = 29 + \boxed{} + \boxed{}$

$\qquad = 30 + \boxed{}$

$\qquad = \boxed{}$

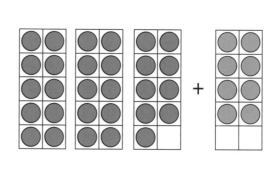

c $38 + 5 = 38 + \boxed{} + \boxed{}$

$\qquad = 40 + \boxed{}$

$\qquad = \boxed{}$

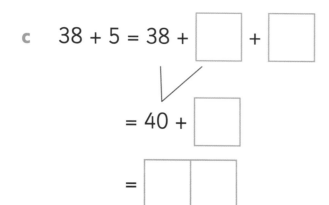

OXFORD UNIVERSITY PRESS

Independent practice

1 Draw getting to a 10 to solve the equations.

a

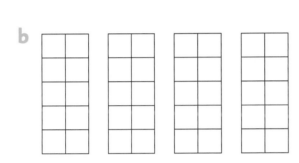

$18 + 9 = \boxed{} + \boxed{} + \boxed{}$

$= \boxed{} + \boxed{}$

$= \boxed{}$

b

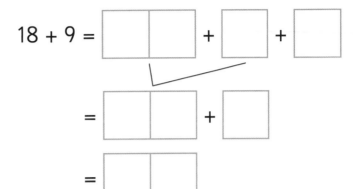

$26 + 7 = \boxed{} + \boxed{} + \boxed{}$

$= \boxed{} + \boxed{}$

$= \boxed{}$

---

2 Solve by getting to a 10.

a $25 + 8 = \boxed{} + \boxed{} + \boxed{}$

$= \boxed{} + \boxed{}$

$= \boxed{}$

b $37 + 6 = \boxed{} + \boxed{} + \boxed{}$

$= \boxed{} + \boxed{}$

$= \boxed{}$

1. Use doubles to help you find the answers.

a   20 + 22 = ☐☐

b   25 + 28 = ☐☐

c   30 + 33 = ☐☐

d   40 + 42 = ☐☐

e   50 + 51 = ☐☐☐

f   55 + 57 = ☐☐☐

2. Use getting to a 10 to solve these in your head.

a   28 + 9 = ☐☐

b   36 + 8 = ☐☐

c   47 + 5 = ☐☐

d   59 + 7 = ☐☐

e   66 + 6 = ☐☐

f   78 + 7 = ☐☐

3. Choose which method you will use, then solve the equations.

a   39 + 7 = ☐☐

Near doubles	Getting to a 10

b   40 + 41 = ☐☐

Near doubles	Getting to a 10

c   100 + 102 = ☐☐☐

Near doubles	Getting to a 10

OXFORD UNIVERSITY PRESS

Adding on a number line

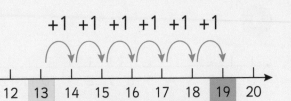

$$13 + 6 = 19$$

Guided practice

Start from the bigger number when you are adding on the number line!

Find the answers with the number lines.

1

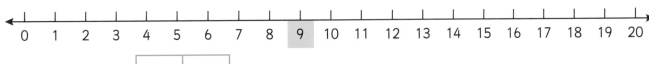

9 + 5 =

2

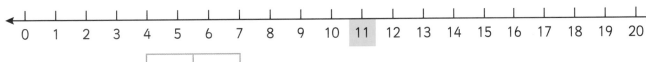

11 + 7 =

3

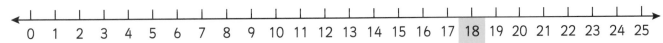

6 + 18 =

4

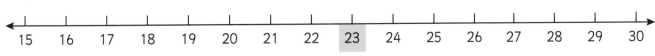

23 + 5 =

Why is it easier to start from the bigger number?

Use the number lines to find the answers.

1

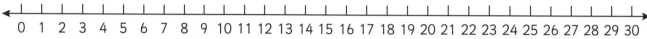

0 1 2 3 4 5 6 7 8 9 10 11 12 13 14 15 16 17 18 19 20 21 22 23 24 25 26 27 28 29 30

19 + 8 = ☐☐

2

15 16 17 18 19 20 21 22 23 24 25 26 27 28 29 30 31 32 33 34 35

24 + 6 = ☐☐

3

0 1 2 3 4 5 6 7 8 9 10 11 12 13 14 15 16 17 18 19 20 21 22 23 24 25 26 27 28 29 30

7 + 14 = ☐☐

4

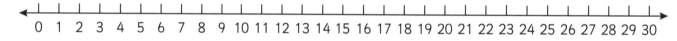

0 1 2 3 4 5 6 7 8 9 10 11 12 13 14 15 16 17 18 19 20 21 22 23 24 25 26 27 28 29 30

5 + 21 = ☐☐

5

30 31 32 33 34 35 36 37 38 39 40 41 42 43 44 45 46 47 48 49 50

32 + 10 = ☐☐

OXFORD UNIVERSITY PRESS

Extended practice

1 Show on the number lines and solve.

a

14 + 5 = [|]

b

21 + 6 = [|]

c

32 + 7 = [|]

2 Solve on the number lines.

a

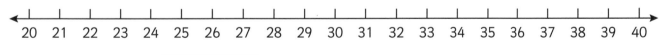

23 + 12 = [|]

b

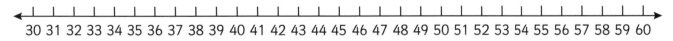

35 + 24 = [|]

You can add numbers in any order.

6 + 3 + 4 = 6 + 4 + 3 = 4 + 3 + 6

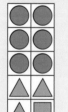 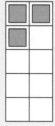 is the same as  is the same as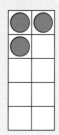

Guided practice

Fill in the gaps.

When might it be useful to change the order of the numbers you are adding?

1 5 + 4 = 4 + ☐ = 9

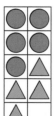 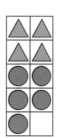

2 7 + 2 + 4 = 7 + ☐ + ☐ = ☐☐

3 4 + 6 + 5 = ☐ + 5 + ☐ = ☐☐

 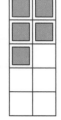 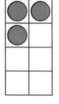

OXFORD UNIVERSITY PRESS

1 Draw and solve.

a 7 + 5 + 1 = ☐ + ☐ + ☐ = ☐☐

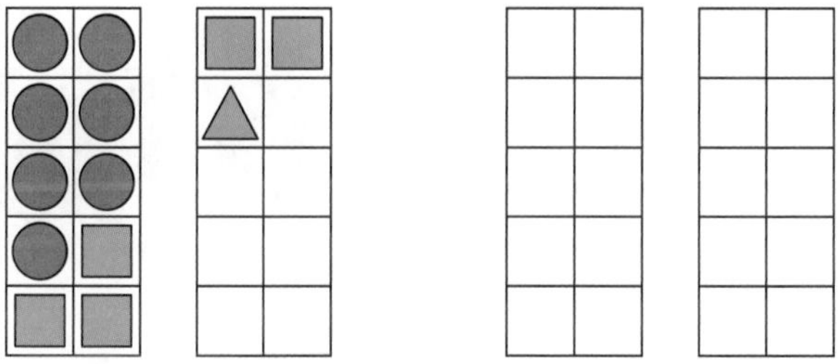

b 2 + 9 + 4 = ☐ + ☐ + ☐ = ☐☐

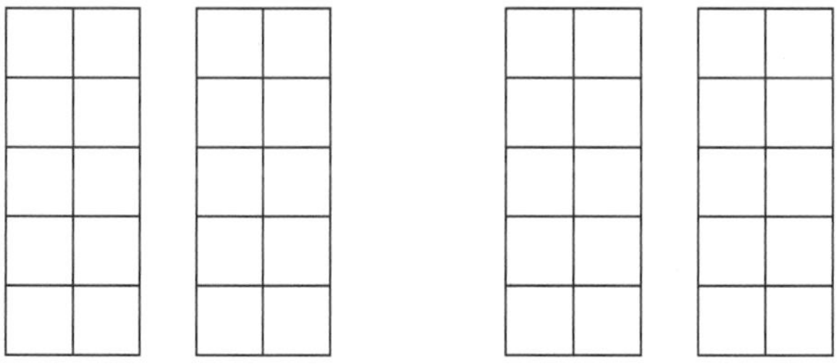

c 8 + 1 + 7 = ☐ + ☐ + ☐ = ☐☐

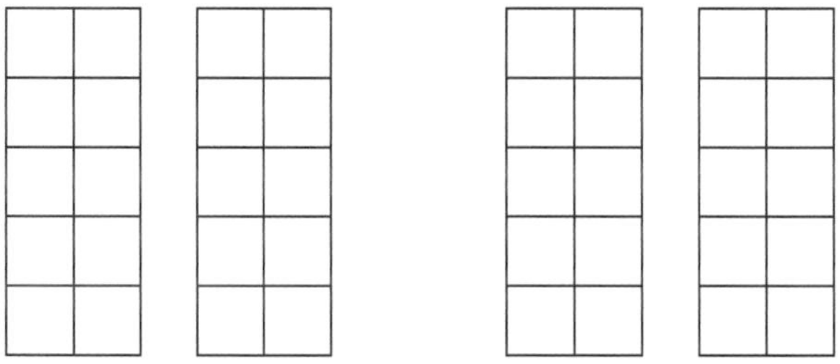

---

2 Fill in the gaps.

a 4 + 7 + 5 = ☐ + ☐ + ☐ = ☐ + ☐ + ☐

= ☐☐

b 1 + 9 + 4 = ☐ + ☐ + ☐ = ☐ + ☐ + ☐

= ☐☐

Extended practice

1 Fill in the missing numbers.

a [ ] + 5 + 7 = [ ] + 8 + 7 = 20

b 6 + 9 + [ ] = 4 + 9 + [ ] = 19

c 8 + 3 + 4 = 4 + [ ] + [ ] = [ ][ ]

d 22 = 9 + 7 + [ ] = [ ] + 7 + 9

e 19 = [ ] + 4 + 7 = 7 + 8 + [ ]

Can you change the order of the numbers when you are doing subtraction?

---

2 Use the numbers to make 3 addition sums.

a 3, 5, 7, 15

[ ] + [ ] + [ ] = [ ]   [ ] + [ ] + [ ] = [ ]

[ ] + [ ] + [ ] = [ ]

b 8, 2, 9, 19

[ ] + [ ] + [ ] = [ ]   [ ] + [ ] + [ ] = [ ]

[ ] + [ ] + [ ] = [ ]

c 9, 21, 7, 5

[ ] + [ ] + [ ] = [ ]   [ ] + [ ] + [ ] = [ ]

[ ] + [ ] + [ ] = [ ]

OXFORD UNIVERSITY PRESS

Getting to a 10

$$24 - 6 = 24 - 4 - 2$$
$$= 20 - 2$$
$$= 18$$

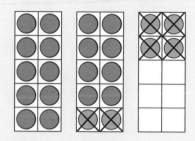

It's easier to subtract numbers from a tens number like 10, 20 or 30.

Guided practice

1. Subtract by getting to a 10.

a   13 − 4 = 13 − $\boxed{3}$ − $\boxed{}$

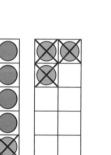

= 10 − $\boxed{}$

= $\boxed{}$

b   21 − 5 = 21 − $\boxed{}$ − $\boxed{}$

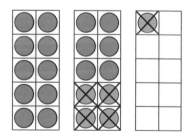

= 20 − $\boxed{}$

= $\boxed{}$

c   32 − 5 = 32 − $\boxed{}$ − $\boxed{}$

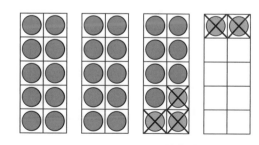

= 30 − $\boxed{}$

= $\boxed{}$

1 Show getting to a 10 to solve the equations.

a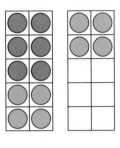

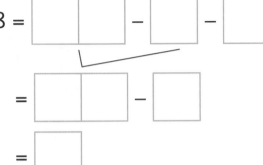

$14 - 8 =$ ☐☐ $-$ ☐ $-$ ☐

$=$ ☐☐ $-$ ☐

$=$ ☐

b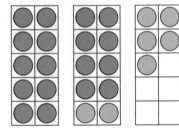

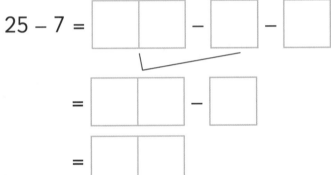

$25 - 7 =$ ☐☐ $-$ ☐ $-$ ☐

$=$ ☐☐ $-$ ☐

$=$ ☐☐

2 Solve by getting to a 10.

a $22 - 6 =$ ☐☐ $-$ ☐ $-$ ☐

$=$ ☐☐ $-$ ☐

$=$ ☐☐

b $35 - 8 =$ ☐☐ $-$ ☐ $-$ ☐

$=$ ☐☐ $-$ ☐

$=$ ☐☐

OXFORD UNIVERSITY PRESS

Counting up to friendly numbers

22 − 18

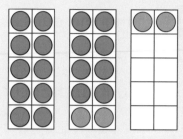

Count up from 18 to 20. $18 + 2 = 20$

Count up from 20 to 22. $20 + 2 = 22$

The difference between 18 and 22 is 4 (2 + 2), so 22 − 18 = 4.

Guided practice

1 Count up to find the answers.

a 13 − 5

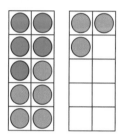

Count up from 5 to 10. $5 + \boxed{} = 10$

Count up from 10 to 13. $10 + \boxed{} = 13$

The difference between 13 and 5

is $\boxed{5}$ + $\boxed{3}$ OR $\boxed{}$.

So 13 − 5 = $\boxed{}$.

b 24 − 19

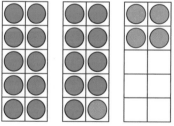

Count up from 19 to 20. $19 + \boxed{} = 20$

Count up from 20 to 24. $20 + \boxed{} = 24$

The difference between 19 and 24

is $\boxed{}$ + $\boxed{}$ OR $\boxed{}$.

So 24 − 19 = $\boxed{}$.

Why do we count up from the smaller number?

Independent practice

1 Fill in the gaps.

a 14 − 8

Count up from ⬜ to ⬜⬜ .

8 + ⬜ = ⬜⬜

Count up from 10 to ⬜⬜ .

10 + ⬜ = ⬜⬜

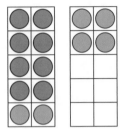

The difference between 14 and 8 is ⬜ + ⬜ OR ⬜ .

So 14 − 8 = ⬜ .

b 23 − 17

17 + ⬜ = ⬜⬜

20 + ⬜ = ⬜⬜

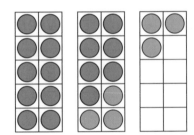

So 23 − 17 = ⬜ .

2 Count up to find the difference.

a

| 0 | 1 | 2 | 3 | 4 | 5 | 6 | 7 | 8 | ⑨ | 10 | 11 | 12 | 13 | 14 | 15 | ⑯ | 17 | 18 | 19 | 20 |

16 − 9 = ⬜

b

0 1 2 3 4 5 6 7 8 9 10 11 12 13 14 15 16 17 18 ⑲20 21 22 23 24 ㉕26 27 28 29 30

25 − 19 = ⬜

OXFORD UNIVERSITY PRESS

Extended practice

1 Use getting to a 10 to solve in your head.

a   12 – 4 = ☐

b   15 – 8 = ☐

c   21 – 9 = ☐☐

d   32 – 6 = ☐☐

e   46 – 7 = ☐☐

f   53 – 5 = ☐☐

---

2 Count up to find the answers.

a   18 – 7 = ☐☐

b   22 – 15 = ☐

c   35 – 23 = ☐☐

d   38 – 27 = ☐☐

e   43 – 36 = ☐

f   48 – 29 = ☐☐

---

3 Choose which strategy to use and solve in your head.

a   28 – 16 = ☐☐

Getting to a 10	Counting up

b   34 – 8 = ☐☐

Getting to a 10	Counting up

c   41 – 34 = ☐

Getting to a 10	Counting up

Subtracting on a number line

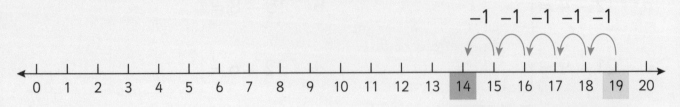

$$19 - 5 = 14$$

Guided practice

Which way do you go on the number line when you are subtracting?

Find the answers with the number lines.

1   14 – 6 =

2   18 – 7 =

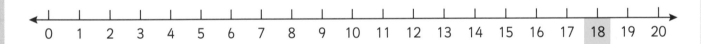

3   23 – 8 =

4   27 – 9 =

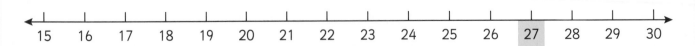

OXFORD UNIVERSITY PRESS

Independent practice

Use the number lines to find the answers.

1 28 – 7 = [|]

0 1 2 3 4 5 6 7 8 9 10 11 12 13 14 15 16 17 18 19 20 21 22 23 24 25 26 27 28 29 30

2 25 – 8 = [|]

0 1 2 3 4 5 6 7 8 9 10 11 12 13 14 15 16 17 18 19 20 21 22 23 24 25 26 27 28 29 30

3 34 – 6 = [|]

15 16 17 18 19 20 21 22 23 24 25 26 27 28 29 30 31 32 33 34 35 36 37 38 39 40

4 43 – 9 = [|]

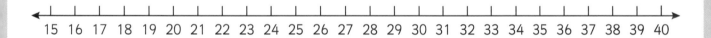

25 26 27 28 29 30 31 32 33 34 35 36 37 38 39 40 41 42 43 44 45 46 47 48 49 50

5 48 – 12 = [|]

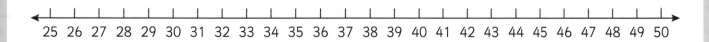

25 26 27 28 29 30 31 32 33 34 35 36 37 38 39 40 41 42 43 44 45 46 47 48 49 50

Extended practice

1 Show on the number lines and solve.

a 19 – 4 = ☐☐

b 28 – 8 = ☐☐

c 33 – 7 = ☐☐

d 36 – 14 = ☐☐

2 Solve on the number lines.

a 37 – 4 – 5 = ☐☐

25 26 27 28 29 30 31 32 33 34 35 36 37 38 39 40 41 42 43 44 45 46 47 48 49 50

b 41 – 7 – 6 = ☐☐

25 26 27 28 29 30 31 32 33 34 35 36 37 38 39 40 41 42 43 44 45 46 47 48 49 50

OXFORD UNIVERSITY PRESS

Addition and subtraction are connected.

If I know 5 + 7 = 12, I also know 12 − 5 = 7 and 12 − 7 = 5.

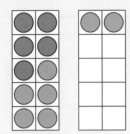 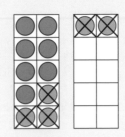 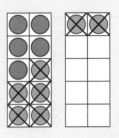

Guided practice

1 Write two subtraction facts to match the addition fact.

Why do both subtraction equations start with the same number?

a 6 + 10 = 16

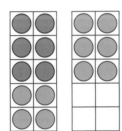

16 − | 1 | 0 | = | 6 |

16 − | 6 | =

b 15 + 4 = 19

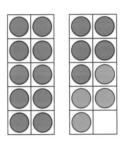

19 − | | = | |

19 − | | = | |

c 9 + 11 = 20

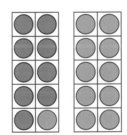

| | − 9 = | |

| | − 11 = | |

d 16 + 7 = 23

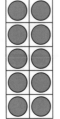 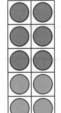 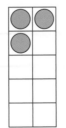

23 − | | = | |

| | − | | = | |

Independent practice

1 Fill in the gaps.

a

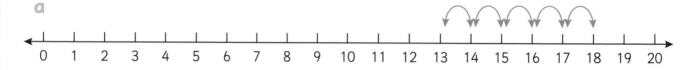

$$13 + 5 = 18$$

$$18 - \boxed{} = \boxed{}$$

$$5 + \boxed{} = 18$$

$$18 - \boxed{} = \boxed{}$$

b

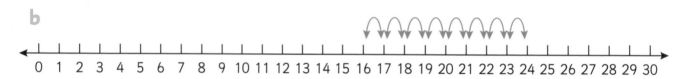

$$16 + \boxed{} = 24$$

$$\boxed{} - 8 = 16$$

$$8 + 16 = 24$$

$$24 - \boxed{} = 8$$

c

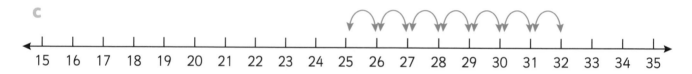

$$25 + \boxed{} = \boxed{}$$

$$\boxed{} - 7 = 25$$

$$7 + \boxed{} = \boxed{}$$

$$\boxed{} - 25 = 7$$

2 Write a matching subtraction fact.

a $8 + 6 = 14$ $\boxed{} - \boxed{} = \boxed{}$

b $16 + 10 = 26$ $\boxed{} - \boxed{} = \boxed{}$

c $13 + 12 = 25$ $\boxed{} - \boxed{} = \boxed{}$

d $11 + 27 = 38$ $\boxed{} - \boxed{} = \boxed{}$

OXFORD UNIVERSITY PRESS

1 Write addition and subtraction facts to match the pictures.

a

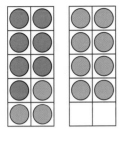

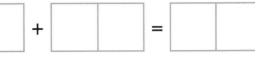

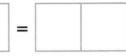

b

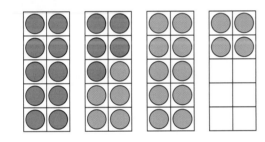

2 Solve on the number lines and use addition to check your answers.

a

0 1 2 3 4 5 6 7 8 9 10 11 12 13 14 15 16 17 18 19 20 21 22 23 24 25 26 27 28 29 30

24 – 5 = ☐☐   ☐☐ + ☐ = 24

b

20 21 22 23 24 25 26 27 28 29 30 31 32 33 34 35 36 37 38 39 40

35 – 12 = ☐☐   ☐☐ + ☐☐ = 35

$$2 + 2 + 2 + 2 = 4 \times 2 = 8$$

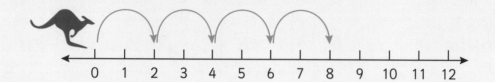

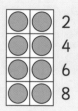

2
4
6
8

Guided practice

Show on the number lines and fill in the blanks.

1 $3 + 3 + 3 = \boxed{} \times \boxed{} = \boxed{}$

Where would I land if I added another 3?

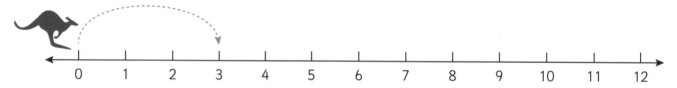

2 $5 + 5 + 5 + 5 + 5 = \boxed{} \times \boxed{} = \boxed{}$

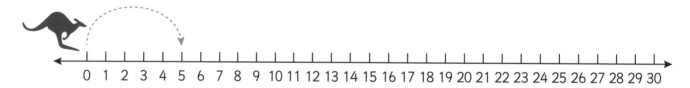

Fill in the blanks.

3

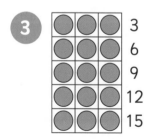

3
6
9
12
15

$\boxed{} + \boxed{} + \boxed{} + \boxed{} + \boxed{} = 15$

$\boxed{}$ threes are $\boxed{}$

$\boxed{} \times \boxed{} = \boxed{}$

OXFORD UNIVERSITY PRESS

Write the equation.

1.

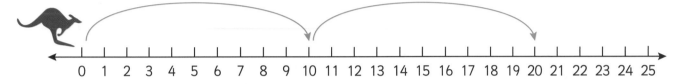

2.

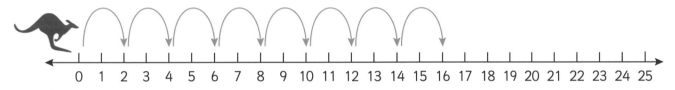

3.

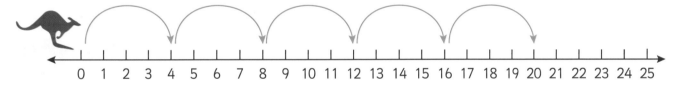

Show on the number line.

4. $3 \times 5 = $ ▢▢

5. $6 \times 2 = $ ▢▢

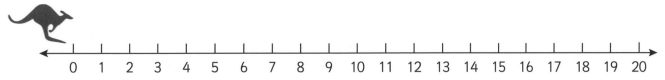

6 7 × 3 = ⬜⬜

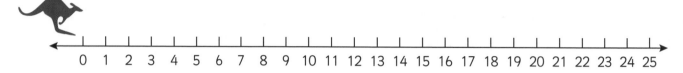

0 1 2 3 4 5 6 7 8 9 10 11 12 13 14 15 16 17 18 19 20 21 22 23 24 25

Write an equation to match the array.

7

⬜ × ⬜ = ⬜⬜

8

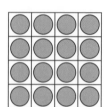

⬜ × ⬜ = ⬜⬜

9

⬜ × ⬜⬜ = ⬜⬜

Draw the array.

I wonder if 4 × 5 is the same as 5 × 4.

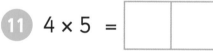

10 3 × 6 = ⬜⬜

11 4 × 5 = ⬜⬜

OXFORD UNIVERSITY PRESS

1

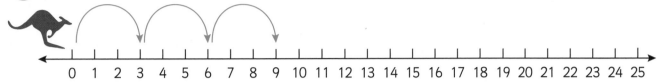

If the kangaroo continues, will she land on … ?

14 | Yes | No   18 | Yes | No   20 | Yes | No

23 | Yes | No   21 | Yes | No   30 | Yes | No

2

How many donuts?   Show how you got your answer.

3 Draw three different arrays of 12 counters.

☐ × ☐ = 12   ☐ × ☐ = 12   ☐ × ☐ = 12

How many groups of 2 in 8?

8 divided into groups of 2 gives 4 groups

OR $8 \div 2 = 4$

$8 - 2 = 6$ $6 - 2 = 4$ $4 - 2 = 2$ $2 - 2 = 0$

What does "equal groups" mean?

Guided practice

1 Draw circles to make equal groups of:

a | 4 |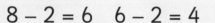

$12 - 4 = \boxed{}$ $\boxed{} - 4 = \boxed{}$ $\boxed{} - 4 = 0$

12 divided by 4 is $\boxed{}$ $12 \div 4 = \boxed{}$

b | 3 |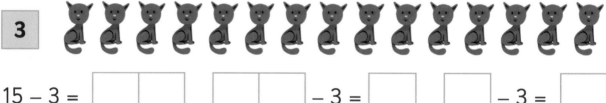

$15 - 3 = \boxed{}\boxed{}$ $\boxed{} - 3 = \boxed{}$ $\boxed{} - 3 = \boxed{}$

$\boxed{} - 3 = \boxed{}$ $\boxed{} - 3 = \boxed{}$

15 divided by 3 is $\boxed{}$ $15 \div 3 = \boxed{}$

2 Equal or unequal shares?

a

| Equal | Unequal |

b

| Equal | Unequal |

OXFORD UNIVERSITY PRESS

Independent practice

1 Share the items equally and fill in the blanks.

a

9 – ☐ = ☐   ☐ – ☐ = ☐   ☐ – ☐ = ☐

9 divided by 3 = ☐   9 ÷ 3 = ☐

b

12 – ☐ = ☐☐   ☐☐ – ☐ = ☐

☐ – ☐ = ☐   6 – ☐ = ☐

☐ – ☐ = ☐   ☐ – ☐ = ☐

12 divided by 6 = ☐   12 ÷ 6 = ☐

c

10 – ☐ = ☐   ☐ – ☐ = ☐

10 divided by 2 = ☐   10 ÷ 2 = ☐

Make the shares equal.

a

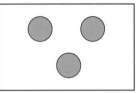  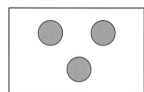 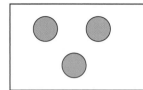

12 ÷ ☐ = ☐

b

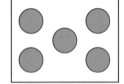 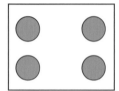 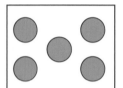 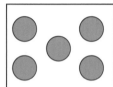 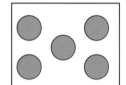

25 ÷ ☐ = ☐

c

     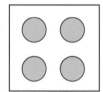

24 ÷ ☐ = ☐

3 Write the equation.

a
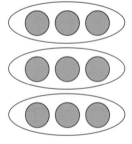

b
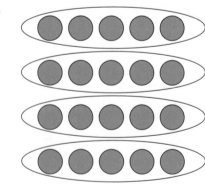

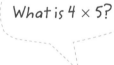
What is 4 × 5?

☐ ÷ ☐ = ☐   ☐☐ ÷ ☐ = ☐

OXFORD UNIVERSITY PRESS

1 Draw 3 ways to equally share 16.

16 ÷ ☐ = ☐   16 ÷ ☐ = ☐   16 ÷ ☐ = ☐

2 There are 4 students in each row.

How many rows of students in a class of:

12? ☐   20? ☐   28? ☐

3 Circle the groups that can be shared equally between 3.

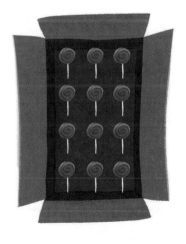  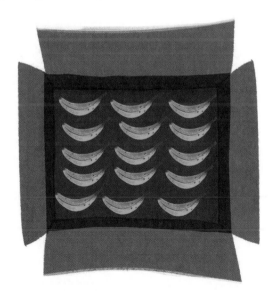

If we remember some things ...   we can work out other things.

I am Sam.

I am 7.

4 + 2 = 6
6 − 2 = 4

4 + 2 = 6
So, 40 + 20 = 60.

6 − 2 = 4
So, 60 − 20 = 40.

60 − 20 = 40
So, 60 − 40 = 20.

Practice helps us to remember.

Guided practice

1 Write and remember these addition facts.

a   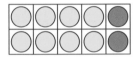    ☐ + ☐ = ☐☐

b   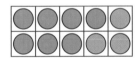 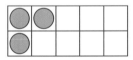   ☐ + ☐ = ☐☐

2 Write and remember these subtraction facts.

a    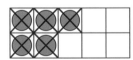   ☐ − ☐ = ☐

b   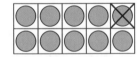 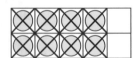   ☐ − ☐ = ☐

3 Write these addition and subtraction facts from memory.

a   7 + 6 = ☐☐   b   8 + 5 = ☐☐   c   15 − 7 = ☐

d   5 + 8 = ☐☐   e   18 − 9 = ☐   f   15 − 8 = ☐

OXFORD UNIVERSITY PRESS

Independent practice

Let's look at some ways to help you learn addition and subtraction facts.

1 Use near doubles.

a 6 + 7 = ?

6 + 7 = 6 + 6 + 1 = 12 + 1 = 13.

7 + 8 = ☐ + ☐ + ☐

= ☐☐ + ☐

= ☐☐

2 Use getting to a 10. ⑦ + 9

7 + 9 = ?   ③   6

10 + 6 = 16

a 9 + 6 = ☐ + ☐ + ☐

= ☐☐ + ☐

= ☐☐

14 – 8 = ?   ⑭ – 8

4   4

10 – 4 = 6

b 17 – 8 = ☐ – ☐ – ☐

= ☐☐ – ☐

= ☐

3 On a separate piece of paper, complete the addition facts as quickly as you can.

1 + 1 =	1 + 2 =	1 + 3 =
1 + 4 =	1 + 5 =	1 + 6 =
1 + 7 =	1 + 8 =	1 + 9 =
2 + 1 =	2 + 2 =	2 + 3 =
2 + 4 =	2 + 5 =	2 + 6 =
2 + 7 =	2 + 8 =	2 + 9 =
3 + 1 =	3 + 2 =	3 + 3 =
3 + 4 =	3 + 5 =	3 + 6 =
3 + 7 =	3 + 8 =	3 + 9 =
4 + 1 =	4 + 2 =	4 + 3 =
4 + 4 =	4 + 5 =	4 + 6 =
4 + 7 =	4 + 8 =	4 + 9 =
5 + 1 =	5 + 2 =	5 + 3 =
5 + 4 =	5 + 5 =	5 + 6 =
5 + 7 =	5 + 8 =	5 + 9 =
6 + 1 =	6 + 2 =	6 + 3 =
6 + 4 =	6 + 5 =	6 + 6 =
6 + 7 =	6 + 8 =	6 + 9 =
7 + 1 =	7 + 2 =	7 + 3 =
7 + 4 =	7 + 5 =	7 + 6 =
7 + 7 =	7 + 8 =	7 + 9 =
8 + 1 =	8 + 2 =	8 + 3 =
8 + 4 =	8 + 5 =	8 + 6 =
8 + 7 =	8 + 8 =	8 + 9 =
9 + 1 =	9 + 2 =	9 + 3 =
9 + 4 =	9 + 5 =	9 + 6 =
9 + 7 =	9 + 8 =	9 + 9 =

4 On a separate piece of paper, complete the subtraction facts as quickly as you can.

10 − 1 =	9 − 1 =	8 − 1 =
7 − 1 =	6 − 1 =	5 − 1 =
4 − 1 =	3 − 1 =	2 − 1 =
11 − 2 =	10 − 2 =	9 − 2 =
8 − 2 =	7 − 2 =	6 − 2 =
5 − 2 =	4 − 2 =	3 − 2 =
12 − 3 =	11 − 3 =	10 − 3 =
9 − 3 =	8 − 3 =	7 − 3 =
6 − 3 =	5 − 3 =	4 − 3 =
13 − 4 =	12 − 4 =	11 − 4 =
10 − 4 =	9 − 4 =	8 − 4 =
7 − 4 =	6 − 4 =	5 − 4 =
14 − 5 =	13 − 5 =	12 − 5 =
11 − 5 =	10 − 5 =	9 − 5 =
8 − 5 =	7 − 5 =	6 − 5 =
15 − 6 =	14 − 6 =	13 − 6 =
12 − 6 =	11 − 6 =	10 − 6 =
9 − 6 =	8 − 6 =	7 − 6 =
16 − 7 =	15 − 7 =	14 − 7 =
13 − 7 =	12 − 7 =	11 − 7 =
10 − 7 =	9 − 7 =	8 − 7 =
17 − 8 =	16 − 8 =	15 − 8 =
14 − 8 =	13 − 8 =	12 − 8 =
11 − 8 =	10 − 8 =	9 − 8 =
18 − 9 =	17 − 9 =	16 − 9 =
15 − 9 =	14 − 9 =	13 − 9 =
12 − 9 =	11 − 9 =	10 − 9 =

5 Circle the more likely answer.

a 48 + 48 = or
$$\begin{matrix} 86 \\ 96 \end{matrix}$$

b 73 − 31 = or
$$\begin{matrix} 42 \\ 41 \end{matrix}$$

c 47 + 27 = or
$$\begin{matrix} 74 \\ 64 \end{matrix}$$

d 89 − 51 = or
$$\begin{matrix} 48 \\ 38 \end{matrix}$$

6 This is Jasmin's homework. Explain what went wrong then write the correct answer.

a 17 − 9 = 26 ✗

b 14 + 15 = 28 ✗

c 35 − 8 = 28 ✗

OXFORD UNIVERSITY PRESS

1 Billy has 50 marbles in one bag and 30 in another. Tilly has 48 marbles in one bag and 48 in another.

a Who has more?

b How do you know?

2 This sentence has 25 letters:

I like to have a birthday party.

Write a different sentence that has 25 letters.

Draw a picture to go with it.

3 Enter 710 on a calculator. Turn it upside down.

Can you see the word "oil"?

a What do you need to add to 350 to get the word "oil"?

b What can you eat if you subtract 37 from 700?

c Subtract 62 from 400 to find a busy insect.

d Find other 3-digit numbers that make calculator words.

This is a whole.	This is a half.	This is a quarter.	This is an eighth.
			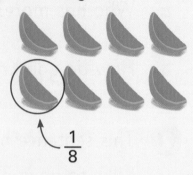
1	$\frac{1}{2}$	$\frac{1}{4}$	$\frac{1}{8}$

Guided practice

1 Draw a square around the whole objects.

2 Circle the fractions.

3 Colour $\frac{1}{2}$ of each shape.

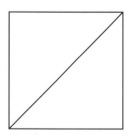

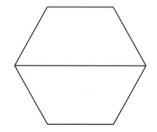

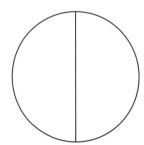

How many halves make a whole?

OXFORD UNIVERSITY PRESS

Independent practice

1 What fraction has been shaded?

a 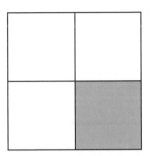 ____

b 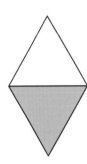 ____

c 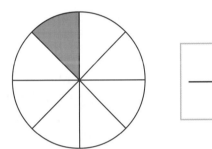 ____

d 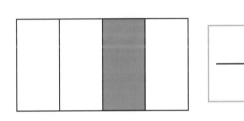 ____

2 Write the size of each piece as a fraction.

a

b

c

____   ____   ____

3 Look at question 2. Which fraction is the biggest? ____

4 Look at question 2. Which fraction is the smallest? ____

5 Label the fractions.

a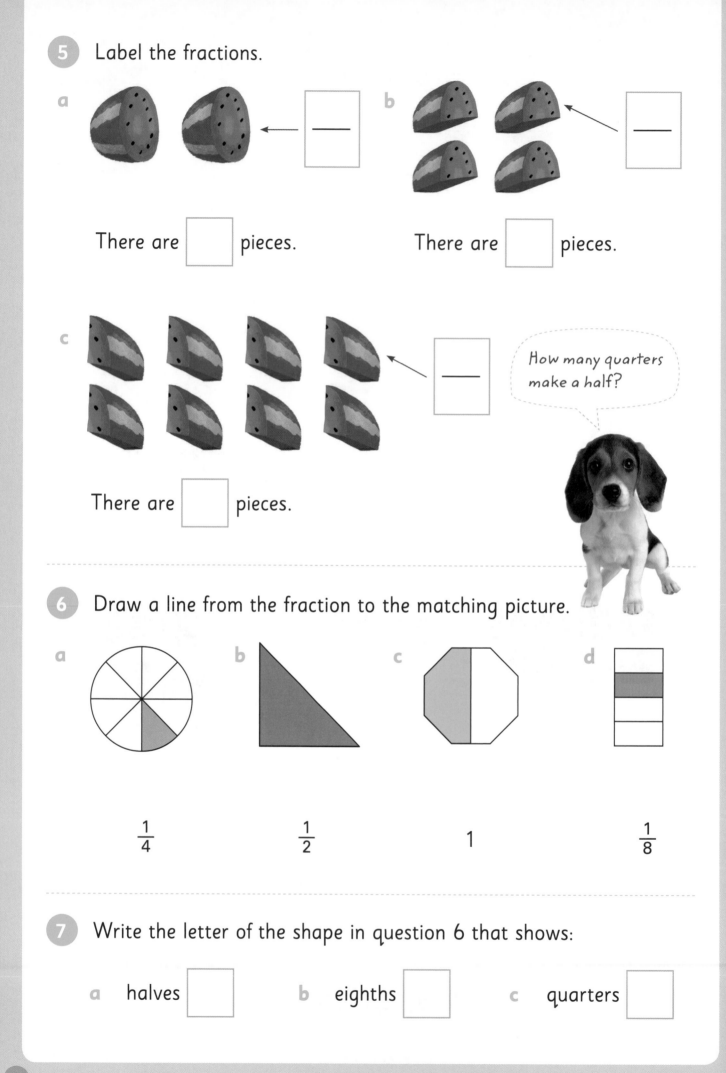

There are ☐ pieces.

b

There are ☐ pieces.

c

There are ☐ pieces.

How many quarters make a half?

6 Draw a line from the fraction to the matching picture.

a

b

c

d

$\frac{1}{4}$

$\frac{1}{2}$

1

$\frac{1}{8}$

7 Write the letter of the shape in question 6 that shows:

a halves ☐

b eighths ☐

c quarters ☐

OXFORD UNIVERSITY PRESS

1 Circle the pictures that show quarters.

a

b

c

d

e

2

a Shade $\frac{1}{2}$ purple.

b Shade $\frac{1}{8}$ red.

c Shade $\frac{1}{4}$ green.

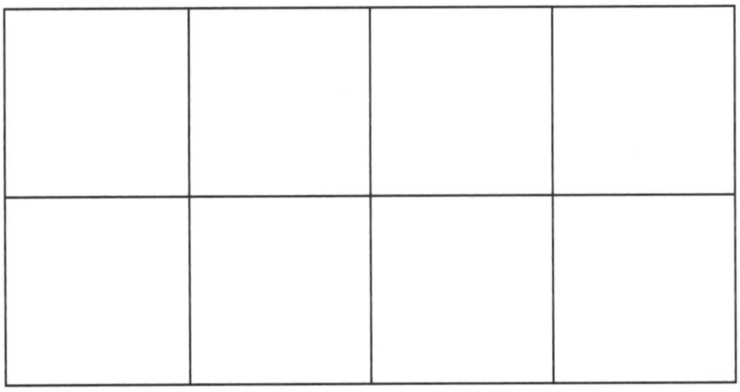

3 Divide into:

a halves.

b quarters.

c eighths.

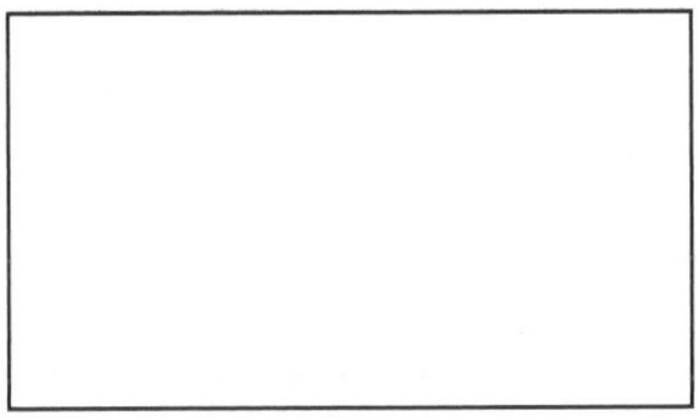

whole 1	half $\frac{1}{2}$	quarter $\frac{1}{4}$	eighth $\frac{1}{8}$

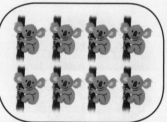   

Is $\frac{1}{2}$, $\frac{1}{4}$ or $\frac{1}{8}$ bigger?

Guided practice

1 Colour $\frac{1}{2}$ of each group.

a

b

2 Colour $\frac{1}{4}$ of each group.

a

b

3 Colour $\frac{1}{8}$ of each group.

a

b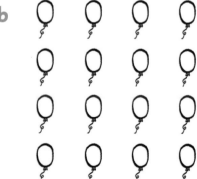

OXFORD UNIVERSITY PRESS

1 What fraction is this?

a

____

b

____

c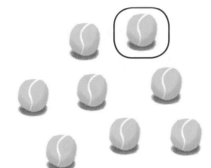

____

d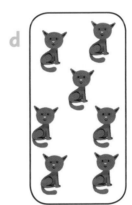

____

2 What fraction of the group are ...

a the yellow blocks?

____

b the red blocks?

____

c the blue blocks?

____

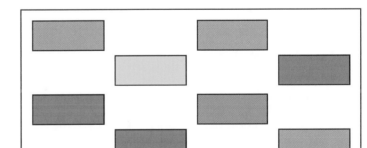

3 **a** Circle $\frac{1}{2}$.

$\frac{1}{2}$ of 4 is ☐ .

b Circle $\frac{1}{4}$.

$\frac{1}{4}$ of 8 is ☐ .

c Circle $\frac{1}{8}$.

$\frac{1}{8}$ of ☐☐ is ☐ .

d Circle $\frac{1}{4}$.

$\frac{1}{4}$ of ☐☐ is ☐ .

4

a Colour $\frac{1}{2}$ red.

b Colour $\frac{1}{4}$ blue.

c Colour $\frac{1}{8}$ green.

What does the number on the bottom of the fraction mean?

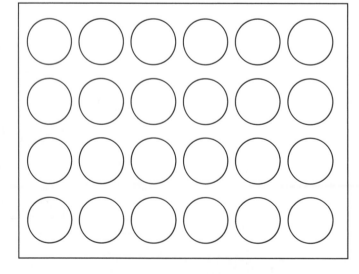

OXFORD UNIVERSITY PRESS

1 Circle which is bigger.

a $\frac{1}{2}$ of 20   OR $\frac{1}{4}$ of 16

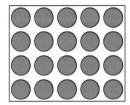

b $\frac{1}{8}$ of 24   OR $\frac{1}{2}$ of 2

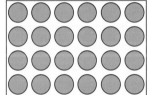

c $\frac{1}{4}$ of 16   OR $\frac{1}{8}$ of 16   OR $\frac{1}{2}$ of 16

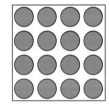

2 What fraction of the group are ...

a girls?

b wearing hats?

c girls AND wearing hats?

To make $1, you can use:

 OR  OR  OR

How else can you make $1?

Guided practice

1 Complete the table.

How many of these ...	... do you need to make this?	Draw the answer	Write the answer
5 CENTS	10 CENTS		5c + 5c = 10c
50 CENTS	2 DOLLARS		
20 CENTS	1 DOLLAR		

2 Complete the table.

How many of these ...	... do you need to make this?	Draw the answer	Write the answer
10 DOLLARS	50 DOLLARS		
50 DOLLARS	100 DOLLARS		
5 DOLLARS	20 DOLLARS		

OXFORD UNIVERSITY PRESS

1 Circle coins that equal:

a $1

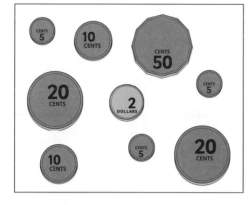

b $2

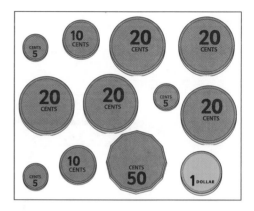

c 20c

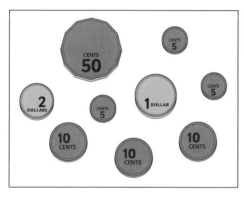

d 75c

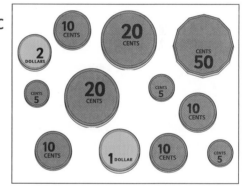

2 Circle notes that equal:

a $20

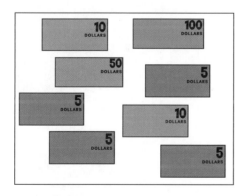

b $50

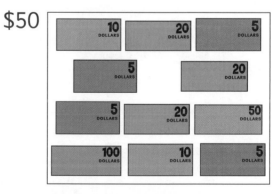

c $100

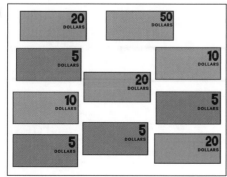

d $45

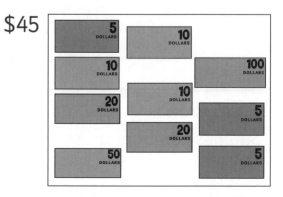

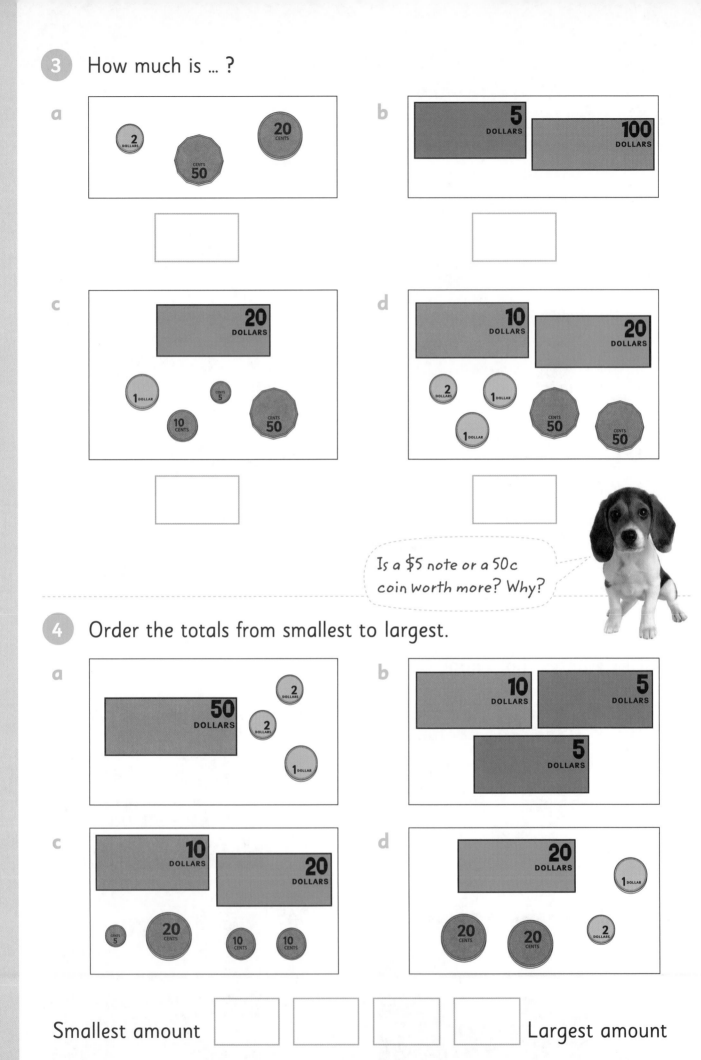

3 How much is … ?

a

b

c

d

Is a $5 note or a 50c coin worth more? Why?

4 Order the totals from smallest to largest.

a

b

c

d

Smallest amount ☐ ☐ ☐ ☐ Largest amount

OXFORD UNIVERSITY PRESS

Extended practice

1 Using the coins and notes we have looked at in this topic, draw 3 different ways to make:

a $2

b $50

c $25

2

a Find 4 ways to make 20c.

b How many ways can you make 25c?

How much money?

   =

20c + 40c = 60c

Guided practice

Count how much.

I can use skip counting to make it easier.

1

2

3

4

5

6

OXFORD UNIVERSITY PRESS

Independent practice

1. Make 40c with:

a 5c coins

b 10c coins

c 20c coins

2. Make $100 with:

a $10 notes

b $20 notes

c $50 notes

3 Rearrange the coins to make them easier to count.

a

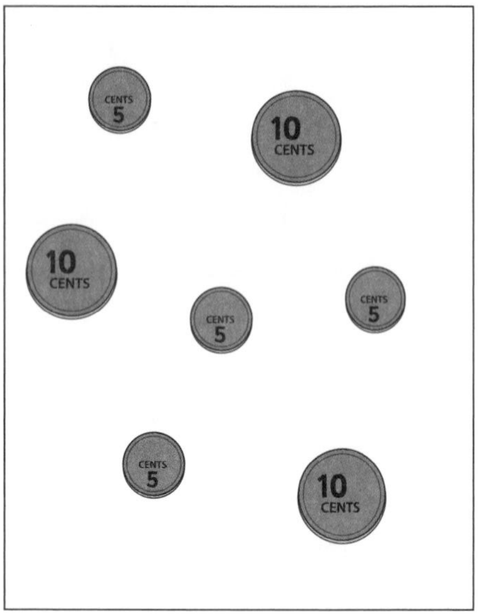

How much?

[ ]

b

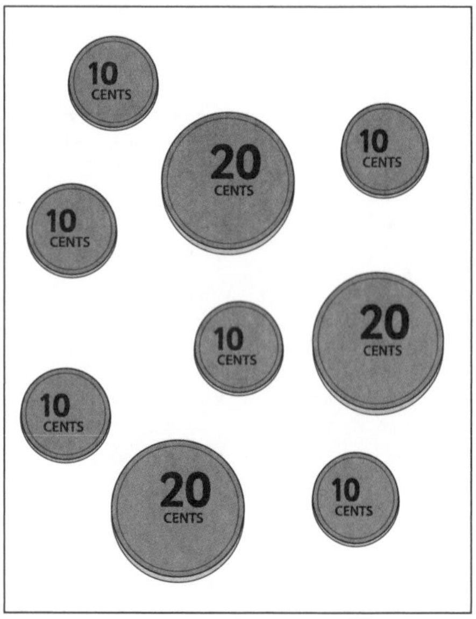

How much?

[ ]

c

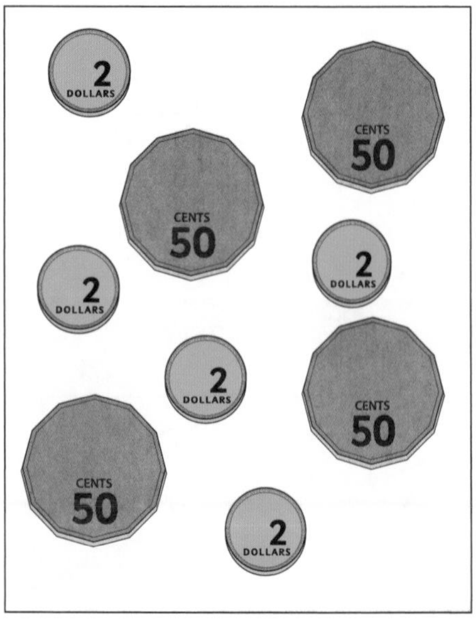

How much?

[ ]

OXFORD UNIVERSITY PRESS

Extended practice

1 Using the coins we have looked at in this topic, draw the least number of coins you could use to make:

a $3.50

Number of coins:

b $6.85

Number of coins:

2 a How much?

b How much would you have left over from this amount if you spent:

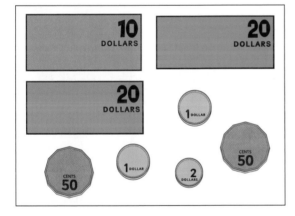

i $20?

ii $45?

iii $32?

3 a How much?

b How much would you have left over from this amount if you spent:

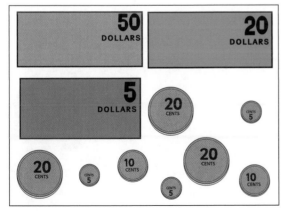

i $20?

ii $45?

iii $32?

Last digit pattern counting by fives: | 5 | , | 0 |

Guided practice

1 Find the last digit pattern, then continue the pattern on the number line.

a Counting by twos: ☐ , ☐ , ☐ , ☐ , ☐

---

b Counting by threes: ☐ , ☐ , ☐ , ☐ , ☐ , ☐ ,

☐ , ☐ , ☐ , ☐

---

c Counting by tens: ☐

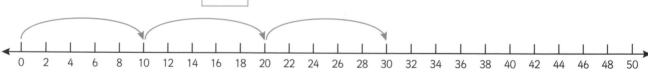

Which number pattern is the longest?

OXFORD UNIVERSITY PRESS

Independent practice

1

a Circle the final digits in the pattern.

4	8	12	16	20	24				

b The pattern is counting by: ☐

c Complete the pattern.

2

a Circle the final digits in the pattern.

80	75	70	65	60	55				

b The pattern is counting by: ☐

c Complete the pattern.

3 Find the missing numbers.

a

	20	30	40		60			90	

b

50		46	44	42			36		

c

4	9	14		24			39		

d

30	27	24			15				

4 Join the dots by following the pattern.

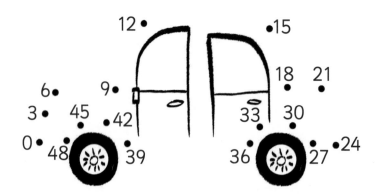

5

0	1	2	3	4	5	6	7	8	9	10
	11	12	13	14	15	16	17	18	19	20
	21	22	23	24	25	26	27	28	29	30
	31	32	33	34	35	36	37	38	39	40
	41	42	43	44	45	46	47	48	49	50
	51	52	53	54	55	56	57	58	59	60
	61	62	63	64	65	66	67	68	69	70
	71	72	73	74	75	76	77	78	79	80
	81	82	83	84	85	86	87	88	89	90
	91	92	93	94	95	96	97	98	99	100

a Circle the numbers counting by 5 from 0.

b Shade the numbers counting by 2 from 0 in yellow.

c Shade the numbers counting by 3 from 0 in red.

6 Which number pattern has the green circle around it?

7 Which numbers are in both the tens and the fives patterns?

8 Which numbers are in the twos, threes, fives and tens patterns?

OXFORD UNIVERSITY PRESS

Extended practice

1

1	2	3	4	5	6	7	8	9	10
11	12	13	14	15	16	17	18	19	20
21	22	23	24	25	26	27	28	29	30
31	32	33	34	35	36	37	38	39	40
41	42	43	44	45	46	47	48	49	50
51	52	53	54	55	56	57	58	59	60
61	62	63	64	65	66	67	68	69	70
71	72	73	74	75	76	77	78	79	80
81	82	83	84	85	86	87	88	89	90
91	92	93	94	95	96	97	98	99	100

a Circle the numbers counting by 5 from 3.

b What is the last digit pattern?

c Colour the numbers counting by 3 from 2.

d What is the last digit pattern?

---

2 What would you be counting by if the last digit pattern was:

a 2, 7, 2, 7, 2, 7?

b 0, 8, 6, 4, 2, 0, 8, 6, 4, 2?

c 7, 7, 7

I wonder if the patterns are going forwards or backwards.

3

a Use these numbers to make a pattern.

37 57 7 27 47 67 17

b What are you counting by?

Word problem

Two monsters went shopping.
They met three more monsters.
How many altogether?

Number sentence

2 + 3 = 5

Guided practice

1 Write a number sentence for the word problems.

a Andrew had 3 cars. He was given 4 more for his birthday. How many did he have altogether?

Number sentence

b Abbey had 8 balloons. But 3 of them popped. How many did she have left?

Number sentence

c There were 9 lemons on the tree. Then 5 new lemons grew. How many lemons now?

Number sentence

Which words show you that it is an addition problem?

OXFORD UNIVERSITY PRESS

Independent practice

1 Draw the problem, then write a number sentence to solve it.

a Tessa has 15 cupcakes. She gives 6 to her friends. How many does she have left?

Picture

Number sentence:

b Hamish has 9 pencils and Primrose has 4 pencils. How many more pencils does Hamish have?

Picture

Number sentence:

c There are 13 candles on the cake. Linus blows out 7 of them. How many are still lit?

Picture

Number sentence:

d Laura read 10 books in April and 6 books in May. How many did she read altogether?

Picture

Number sentence:

2 Write a word problem to match the number sentence.

a   8 + 4 = 12

b   15 − 5 = 10

c   16 + 5 = 21

---

3 Decide if each word problem is addition or subtraction.

a   Remy scored 14 points on Monday and 17 points on Tuesday. What was his total point score?

Addition	Subtraction

b   Jay had 17 marbles. He bought another 12. How many does he have now?

Addition	Subtraction

c   Nina had 16 pairs of shoes. She gave away 14 pairs. How many pairs does she have left?

Addition	Subtraction

OXFORD UNIVERSITY PRESS

Extended practice

1 Write a word problem and number sentence to match each picture.

a

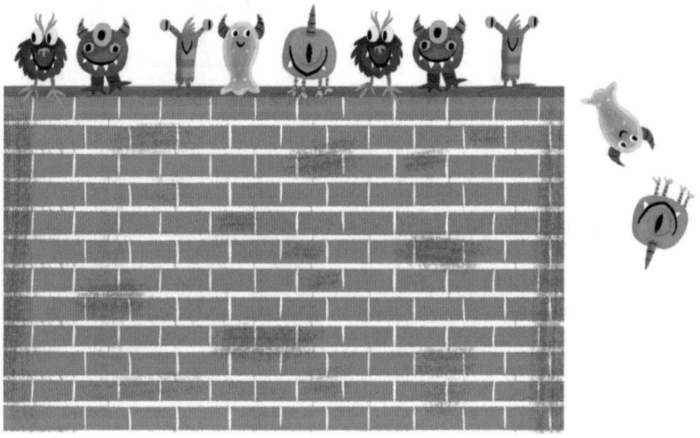

Word problem

Number sentence

b

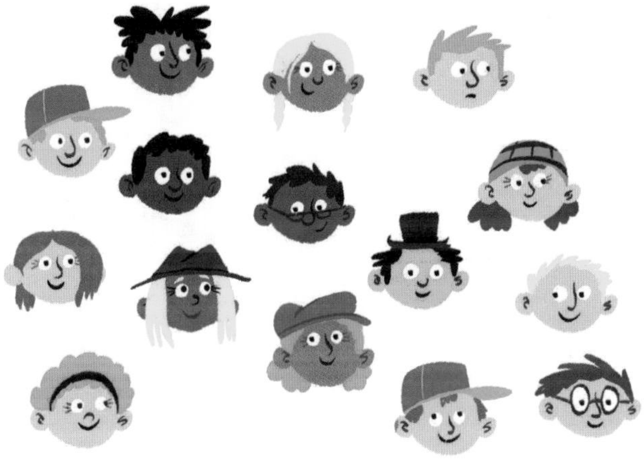

Word problem

Number sentence

c

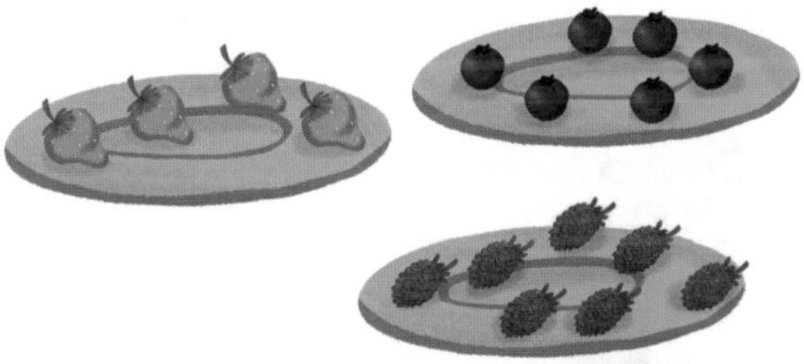

Word problem

Number sentence

When might you need to solve a word problem in real life?

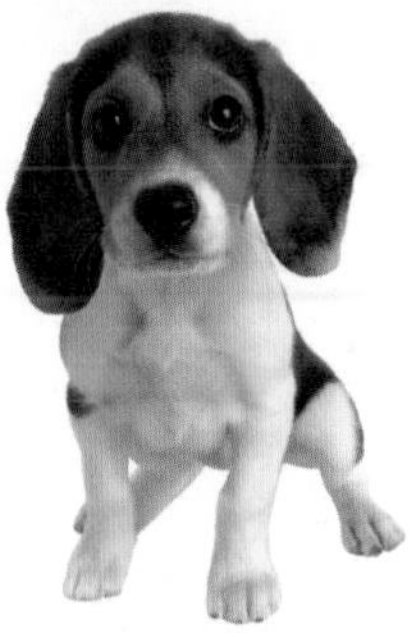

This book is
3 hand spans
long.

This book has
an area of 12
sticky notes.

Why might I need to know
the area of something?

Guided practice

1 Use your hand span to find the length of:

a this book

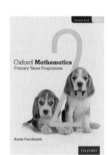

b your table

c the whiteboard

 hand spans

 hand spans

hand spans

2 Use sticky notes to find the area of:

a a book

b your table

c another item

 sticky notes

 sticky notes

sticky notes

OXFORD UNIVERSITY PRESS

Independent practice

1 Complete the table.

Length to find	Unit of length used	Estimate	Actual length
my arm	hand span	4 hand spans	
my pencil			
my chair			
the door			

2 Choose a unit to measure the length of these items.

Unit:

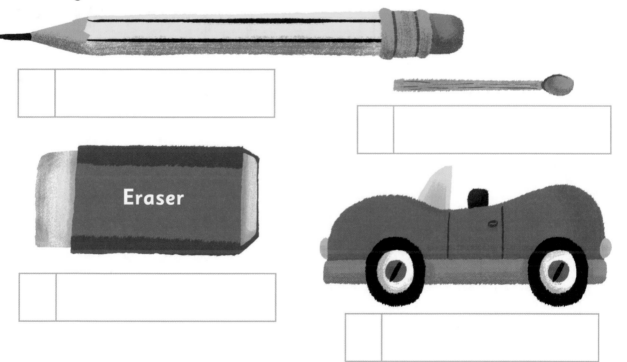

Eraser

3 Order the items from shortest to longest.

pencil		eraser		match		toy car

4 Complete the table.

Area to find	Unit of area used	Estimate	Actual area
my pencil case	erasers	32 erasers	
my writing book			
my lunch box			
my eraser			

5 Find the area of each shape.

a

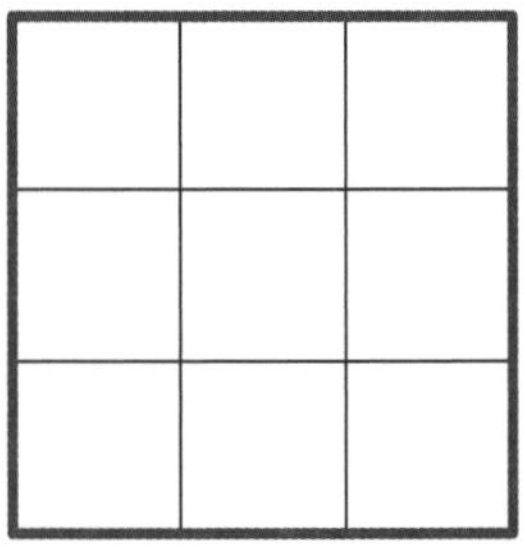

[ ] squares

b

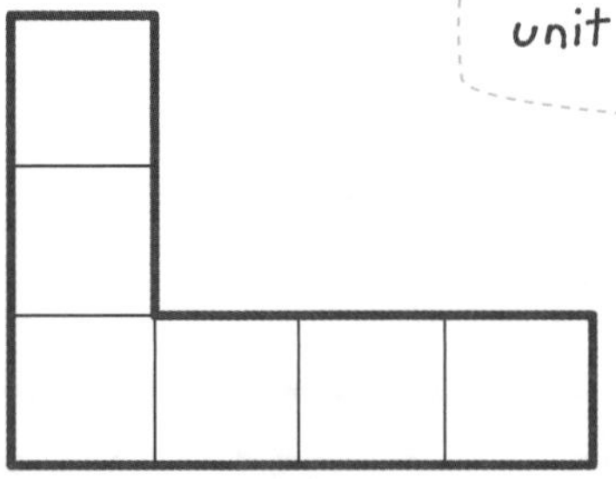

[ ] squares

Would a frisbee be a good unit to measure area with?

c

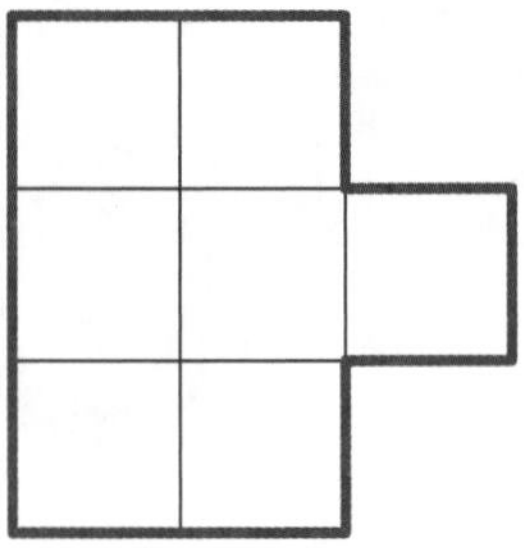

[ ] squares

d

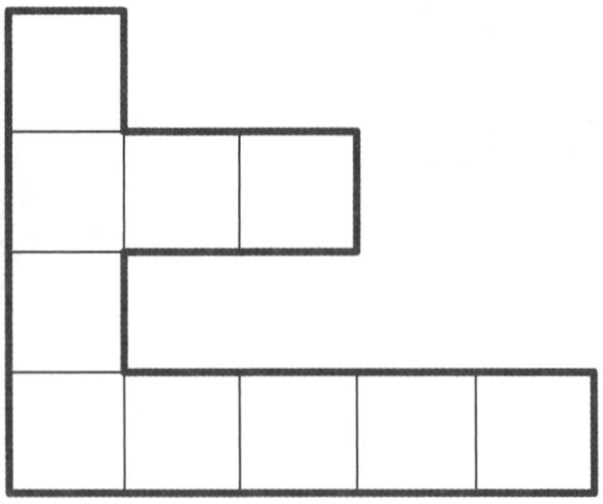

[ ] squares

e

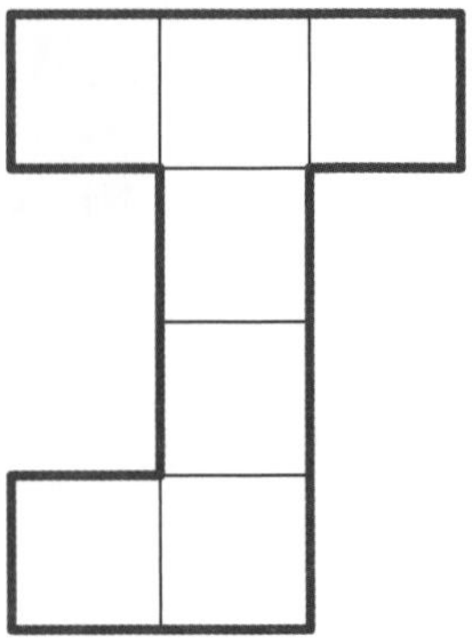

[ ] squares

6 Circle the shape with the largest area.

OXFORD UNIVERSITY PRESS

Extended practice

1 Find the length and area of each shape using units of your choice.

Length	$2\frac{1}{2}$	erasers

Area	9	fingertips

Length		

Area		

Length		

Area		

Length		

Area		

2 Circle the longest shape.

3 Put a tick on the shape with the greatest area.

4 Draw a star on the shape with the smallest area.

Metres

We measure the length of long items in metres (m).

3 m

4 m

What are some things we might measure in metres?

Guided practice

1 Use a metre ruler to find:

a the length of your classroom.

[ ] metres

b the width of a bookcase.

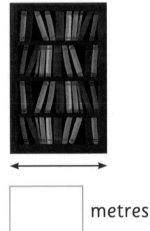

[ ] metres

c the height of the door.

[ ] metres

d the width of the whiteboard.

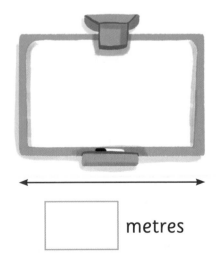

[ ] metres

OXFORD UNIVERSITY PRESS

1. Find items you think fit the estimates, then check with a metre ruler.

Item	Estimate	Actual length
mathematics book	less than 1 metre	less than 1 metre
	less than 1 metre	
	about 1 metre	
	more than 1 metre	

2. Draw lines to match the items with the estimates.

less than 1 metre	about 1 metre	more than 1 metre

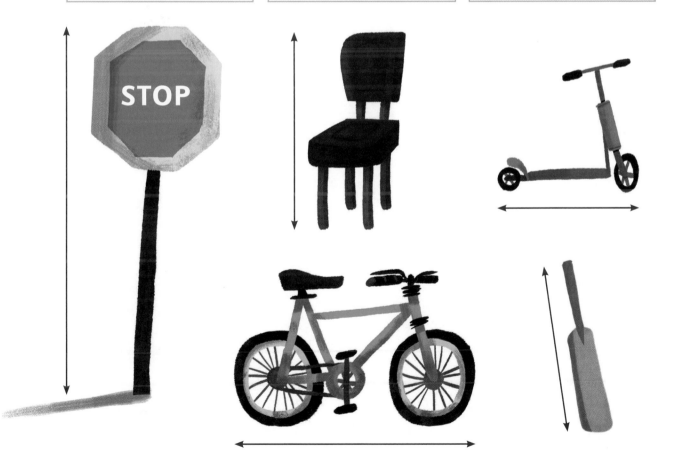

Centimetres

We measure the length of small items in centimetres (cm).

80 cm

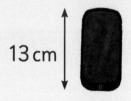

13 cm

30 cm

There are 100 centimetres in 1 metre.

Guided practice

1 Use a 30 cm ruler to find:

a the length of your pencil.

☐ centimetres

b the width of this book.

☐ centimetres

c the width of your hand span.

☐ centimetres

d the length of your pencil case.

☐ centimetres

OXFORD UNIVERSITY PRESS

1 Find items you think fit the estimates, then check with a 30 cm ruler.

Item	Estimate	Actual length
my eraser	less than 30 cm	5 cm
	less than 30 cm	
	about 30 cm	
	more than 30 cm	

2 Draw lines to match the items with the estimates.

less than 30 cm	about 30 cm	more than 30 cm

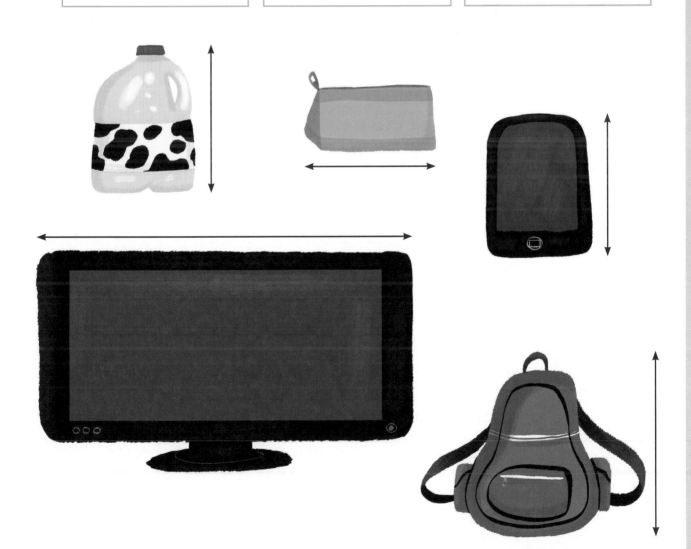

1 Would you measure these items in metres or centimetres?

a

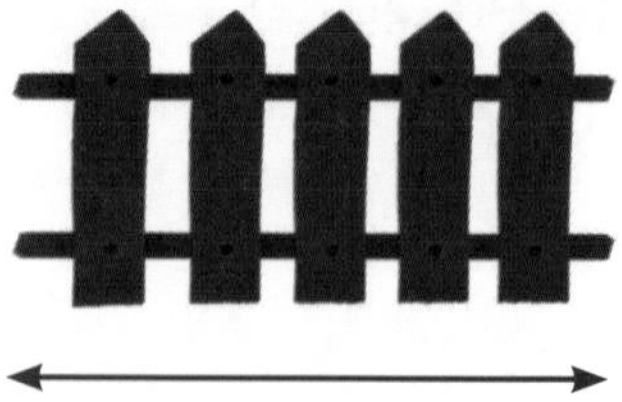

m	cm

b

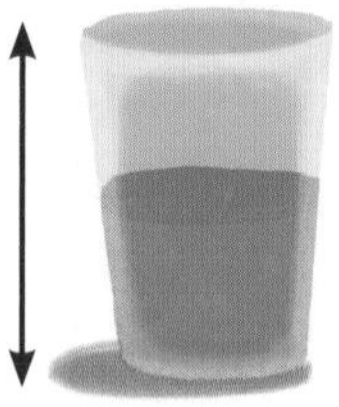

m	cm

c

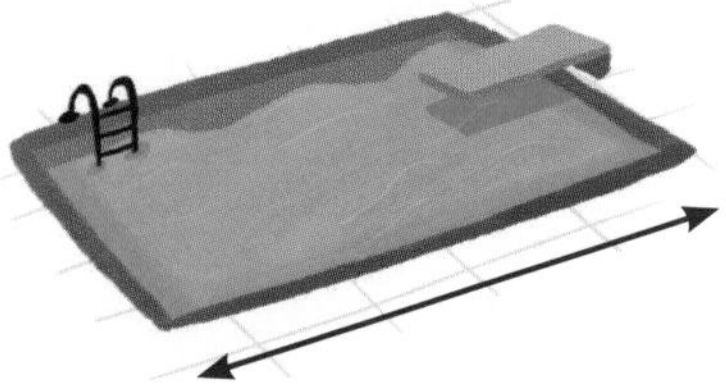

m	cm

d

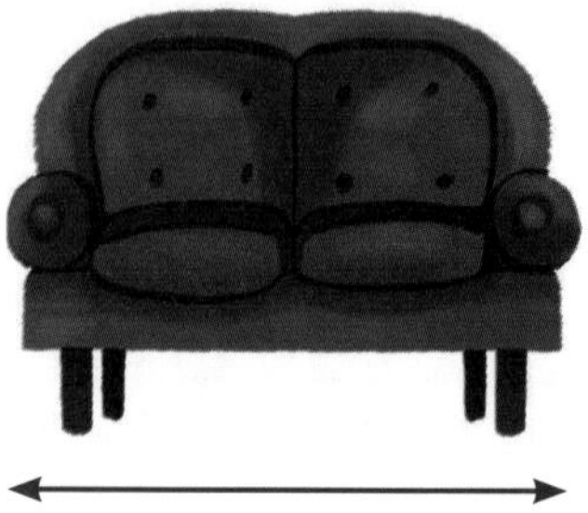

m	cm

e

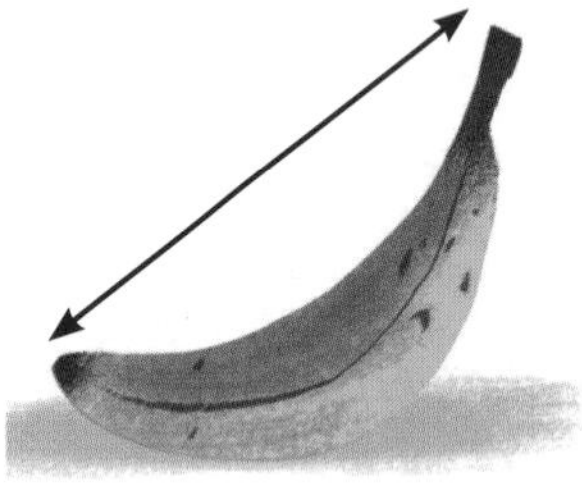

m	cm

f

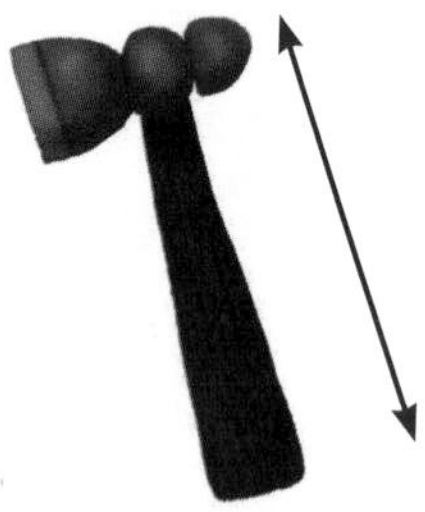

m	cm

2 Which of these items do you estimate is the longest?

3 Which is the shortest?

4 How long might the couch be?

5 How long might the banana be?

OXFORD UNIVERSITY PRESS

The volume of this object is 6 blocks.

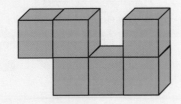

The capacity of this jug is 4 cups.

Guided practice

1 What is the volume of each object?

a

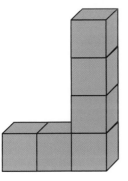

b

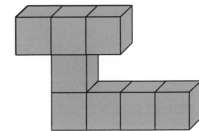

☐ blocks

☐ blocks

2 Tick to estimate the capacity of each container.

a

☐ more than 2 coffee cups

☐ less than 2 coffee cups

b

☐ more than 2 coffee cups

☐ less than 2 coffee cups

Volume is how much space an object takes up. Capacity is how much a container holds.

1

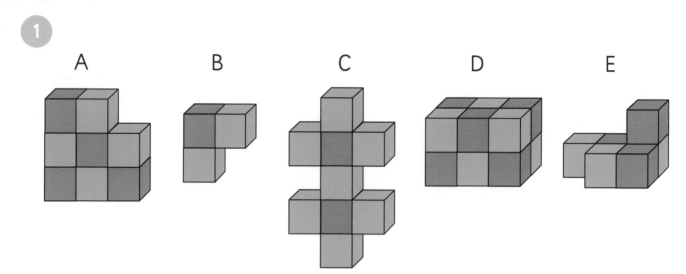

A   B   C   D   E

a   Write the letters to order by volume.

Smallest volume  Largest volume

b   Which objects have a bigger volume than E?

c   Which objects have a smaller volume than C?

2

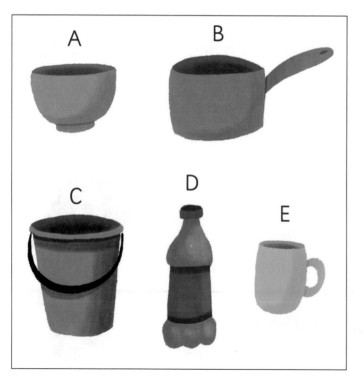

A   B

C   D   E

a   Which container has the biggest capacity?

b   Which container has the smallest capacity?

c   Which containers hold more than D?

d   Which containers hold less than B?

OXFORD UNIVERSITY PRESS

3 Circle the objects with a volume of:

a 8 cubes.

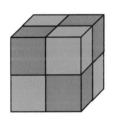 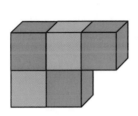 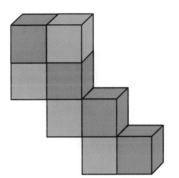 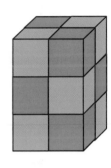

b 10 cubes.

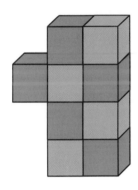 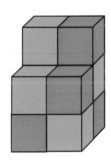 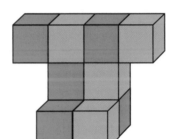 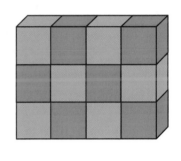

4 Number the items in each group 1, 2, 3 from smallest to largest capacity.

a

☐ ☐ ☐

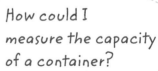
How could I measure the capacity of a container?

b

☐ ☐ ☐

1 Find 1 jug and 4 smaller containers.

   a Estimate how many of each container will fill the jug.

   b Check and record the results.

Container	Estimate	Capacity

2 **a** Build 3 different models using 8 blocks.

   b Draw your models.

3 Select whether each picture is showing capacity or volume.

a

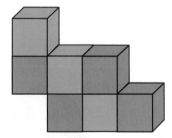

Capacity
Volume

is 7 blocks.

b

Capacity
Volume

is 3 cups.

OXFORD UNIVERSITY PRESS

heavier   lighter   equal

The dog has a greater mass than the teddy.

The apple and the pear have the same mass.

Guided practice

I wonder if bigger objects are always heavier.

1 Circle the **heavier** object.

a   b   c

2 Circle the object with the **smaller** mass.

a   b   c

1 Order by mass.

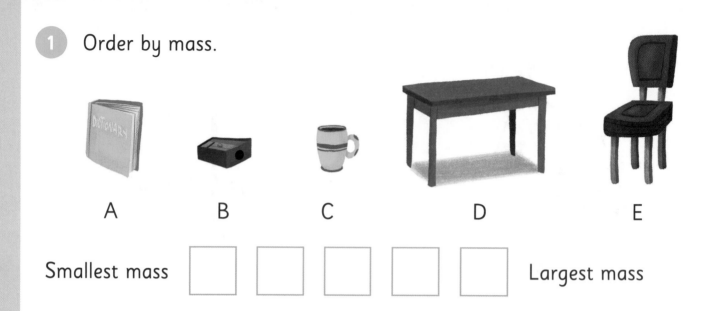

A   B   C   D   E

Smallest mass ☐ ☐ ☐ ☐ ☐ Largest mass

2 Draw the objects at the correct end of the balance scale.

a   b

c   d

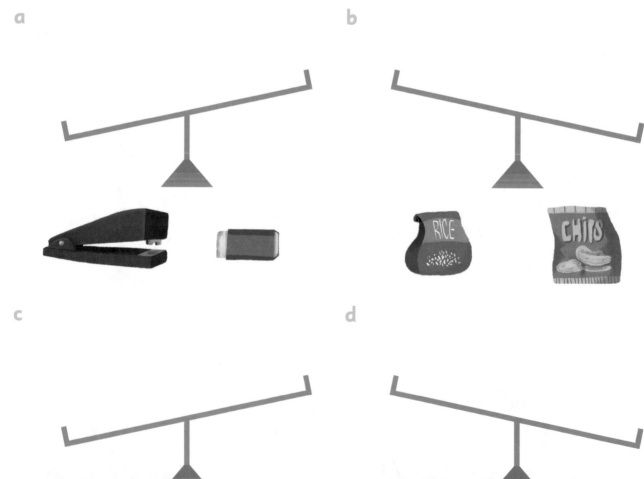

OXFORD UNIVERSITY PRESS

3　Choose 3 pairs of objects.

a　Record the objects in the table.

b　Estimate which item in the pair is heavier.

c　Use a balance scale to check. Record the answer.

How else can you measure the mass of something?

Item 1	Item 2	I estimate this item will be heavier:	The item with the greater mass was:

4　Get a balance scale and a large handful of counters.

a　Choose objects that you think will have lighter, heavier and about the same mass as the counters.

b　Check with the balance scale.

c　Record the results.

		Result
I think this item will be lighter.		
I think this item will be heavier.		
I think this item will be about the same.		

1

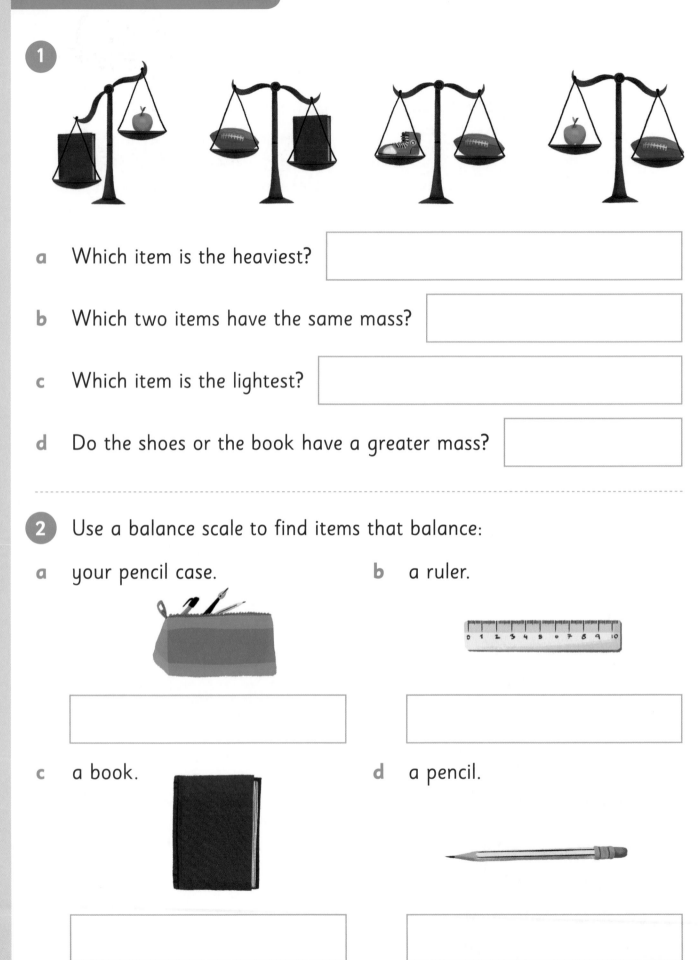

a Which item is the heaviest?

b Which two items have the same mass?

c Which item is the lightest?

d Do the shoes or the book have a greater mass?

2 Use a balance scale to find items that balance:

a your pencil case.

b a ruler.

c a book.

d a pencil.

OXFORD UNIVERSITY PRESS

2 o'clock

quarter past 2

half past 2

quarter to 3

Why do we say "quarter past" and "quarter to"?

Guided practice

1 What number is the minute hand pointing to at:

a

b

c

d

5 o'clock?   quarter to 6?   quarter past 11?   half past 8?

2 Draw a line to match the clocks to the times.

a

b

c

d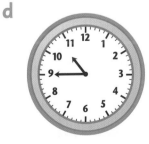

quarter to 11   half past 4   9 o'clock   quarter past 6

1 Draw in the minute hands.

a half past 3

b quarter past 1

c 7 o'clock

d quarter past 12

e quarter to 7

f half past 6

2 Draw in the hour hands.

a quarter past 3

b 8 o'clock

c quarter to 10

d half past 5

e quarter to 9

f 12 o'clock

Why do you think it is called the "hour hand"?

OXFORD UNIVERSITY PRESS

3 Draw the times on the clocks.

a quarter to 5

b quarter past 8

c half past 1

d quarter past 10

e 6 o'clock

f half past 11

What's another way of saying "half past"?

4 Write in the times.

a

b

c

d

e

f

1 What time will it be in:

a 1 hour?

b half an hour?

c quarter of an hour?

d 2 hours?

2 Fill in the missing pieces.

a

2:15

[ ]

b

[:]

3 o'clock

c

3:45

quarter to [ ]

d

12:00

[ ]

e

[: 15]

quarter past 7

f

[:]

[ ]

OXFORD UNIVERSITY PRESS

How long does it take to write the alphabet?

Sam:
35 seconds

Alex:
45 seconds

Gina:
30 seconds

What could you do in **one** second?

Guided practice

1 Look at the picture above.

a Who was the quickest?

b Who took the longest?

c Who was five seconds slower than Gina?

2 There are 60 seconds in 1 minute. It takes 1 minute for the second hand to go right round the clock.

How many minutes in:

a 120 seconds? b 180 seconds?

3 There are 60 minutes in 1 hour. It takes 1 hour for the minute hand to go right round the clock.

60 minutes

How many minutes in:

a 2 hours?

b half an hour?

Independent practice

1 There are 24 hours in a day.

a 2 days = ☐☐ hours

b If you sleep 10 hours in one day, how many hours are you awake? ☐☐

---

2 There are 7 days in a week.

a 14 days = ☐☐ weeks

b 4 weeks = ☐☐ days

c 21 days = ☐☐ weeks

d 10 weeks = ☐☐ days

February is the only month with exactly four weeks, expect in a leap year when it has an extra day!

---

3 A month is about 4 weeks.

a 2 months = about ☐☐ weeks

b March has ☐ weeks and ☐☐ days.

---

4 There are 12 months in a year.
There are 52 weeks in a year.

a 48 months = ☐ years

b 3 years = ☐☐ months

c 104 weeks = ☐ years

Your birthday occurs every 12 months, or 52 weeks.

OXFORD UNIVERSITY PRESS

5 Draw lines to match the times to the events.

Time	Event
A few seconds	Sleeping at night 
A few minutes	A football season 
A few hours	Eating a sandwich 
A few days	Becoming a top athlete 
A few weeks	Writing your name 
A few months	Reading a chapter book 
A few years	The school summer holidays 

6 What takes a few minutes to do at school? Draw and write it.

Extended practice

You will need a stopwatch for the activities on this page.

1   When the second hand moves from one number to the next on a clock, 5 seconds have passed.

Skip count to find the number of seconds that have passed from the top number to:

a   number 2.   b   number 4.

c   the bottom number.   d   number 10.

2   With a partner, try to guess when 10 seconds has passed by following the steps below.

·   Close your eyes.

·   Using a stopwatch, your partner will tell you when to start counting.

·   Count 10 seconds in your mind.

·   Raise your hand when you have finished counting to 10.

a   Did you guess 10 seconds?

b   Try again. Were you better this time?

3   With a partner, time each other writing the alphabet by following the steps below.

·   Using a stopwatch, your partner will tell you when to start counting.

·   Neatly write the alphabet.

·   Raise your hand when you have finished.

·   Your partner will tell you how many seconds it took.

·   Swap roles with your partner and repeat.

How many seconds did it take you?

OXFORD UNIVERSITY PRESS

Southern hemisphere seasons:

summer	autumn	winter	spring

December	March	June	September
January	April	July	October
February	May	August	November

Northern hemisphere seasons:

summer	autumn	winter	spring

June	September	December	March
July	October	January	April
August	November	February	May

Guided practice

1 a How many months are in a year?

b How many seasons are in a year?

c How many months are in a season?

Which hemisphere do you live in? What are the names of the seasons where you live? When do they occur?

a Write the months of the year in order.

b Write the season each month is in for both the northern and southern hemispheres.

Month	Seasons in the southern hemisphere	Seasons in the northern hemisphere
January		

OXFORD UNIVERSITY PRESS

2 **a** Write down the birthdays of 10 people in your class.

Name	Birthday

What month is your birthday in? What season is it in?

b Write the 10 birthdays in the order they occur in the year.

c Write the season each birthday is in.

Name	Birthday	Season

Extended practice

1 In the south of Australia, some Aboriginal people have 6 seasons.

Aboriginal season	High summer	Late summer	Early winter	Deep winter	Early spring	True spring
Months	November December January	February March	March April May	May June July	July August	September October November

a Which of the 4 seasons is not in the Aboriginal season names?

b Which seasons are the shortest?

c How many months are in deep winter?

d Is each Aboriginal season described above in summer, winter, autumn or spring?

Aboriginal season	Summer, winter, autumn or spring?
High summer	
Late summer	
Early winter	
Deep winter	
Early spring	
True spring	

OXFORD UNIVERSITY PRESS

January						
Sun	Mon	Tues	Wed	Thur	Fri	Sat
	1	2	3	4	5	6
7	8	9	10	11	12	13
14	15	16	17	18	19	20
21	22	23	24	25	26	27
28	29	30	31			

The first day of January was a Monday.

The first Sunday in January was the 7th.

The last day of January was a Wednesday.

Guided practice

What do people use calendars for?

1

February						
Sun	Mon	Tues	Wed	Thur	Fri	Sat
		1	2	3	4	5
6	7	8	9	10	11	12
13	14	15	16	17	18	19
20	21	22	23	24	25	26
27	28					

a What is the first day of February?

b What date is the first Sunday in February?

c What is the last day in February?

2

November						
Sun	Mon	Tues	Wed	Thur	Fri	Sat
					1	2
3	4	5	6	7	8	9
10	11	12	13	14	15	16
17	18	19	20	21	22	23
24	25	26	27	28	29	30

a How many Sundays are in November?

b How many Saturdays?

c What day is the 13th of November?

d What date is the last day in November?

Independent practice

1

JANUARY

Sun	Mon	Tues	Wed	Thur	Fri	Sat
	1	2	3	4	5	6
7	8	9	10	11	12	13
14	15	16	17	18	19	20
21	22	23	24	25	26	27
28	29	30	31			

FEBRUARY

Sun	Mon	Tues	Wed	Thur	Fri	Sat
				1	2	3
4	5	6	7	8	9	10
11	12	13	14	15	16	17
18	19	20	21	22	23	24
25	26	27	28			

MARCH

Sun	Mon	Tues	Wed	Thur	Fri	Sat
				1	2	3
4	5	6	7	8	9	10
11	12	13	14	15	16	17
18	19	20	21	22	23	24
25	26	27	28	29	30	31

APRIL

Sun	Mon	Tues	Wed	Thur	Fri	Sat
1	2	3	4	5	6	7
8	9	10	11	12	13	14
15	16	17	18	19	20	21
22	23	24	25	26	27	28
29	30					

MAY

Sun	Mon	Tues	Wed	Thur	Fri	Sat
	1	2	3	4	5
6	7	8	9	10	11	12
13	14	15	16	17	18	19
20	21	22	23	24	25	26
27	28	29	30	31		

JUNE

Sun	Mon	Tues	Wed	Thur	Fri	Sat
					1	2
3	4	5	6	7	8	9
10	11	12	13	14	15	16
17	18	19	20	21	22	23
24	25	26	27	28	29	30

JULY

Sun	Mon	Tues	Wed	Thur	Fri	Sat
1	2	3	4	5	6	7
8	9	10	11	12	13	14
15	16	17	18	19	20	21
22	23	24	25	26	27	28
29	30	31				

AUGUST

Sun	Mon	Tues	Wed	Thur	Fri	Sat
			1	2	3	4
5	6	7	8	9	10	11
12	13	14	15	16	17	18
19	20	21	22	23	24	25
26	27	28	29	30	31	

SEPTEMBER

Sun	Mon	Tues	Wed	Thur	Fri	Sat
						1
2	3	4	5	6	7	8
9	10	11	12	13	14	15
16	17	18	19	20	21	22
23	24	25	26	27	28	29
30						

OCTOBER

Sun	Mon	Tues	Wed	Thur	Fri	Sat
	1	2	3	4	5	6
7	8	9	10	11	12	13
14	15	16	17	18	19	20
21	22	23	24	25	26	27
28	29	30	31			

NOVEMBER

Sun	Mon	Tues	Wed	Thur	Fri	Sat
				1	2	3
4	5	6	7	8	9	10
11	12	13	14	15	16	17
18	19	20	21	22	23	24
25	26	27	28	29	30	

DECEMBER

Sun	Mon	Tues	Wed	Thur	Fri	Sat
						1
2	3	4	5	6	7	8
9	10	11	12	13	14	15
16	17	18	19	20	21	22
23	24	25	26	27	28	29
30	31					

a How many days are in each month?

Month	Days

b On this calendar, which months have 5 Sundays?

c On this calendar, which months start on a Thursday?

OXFORD UNIVERSITY PRESS

2

May						
Sun	Mon	Tues	Wed	Thur	Fri	Sat
			1	2	3	4
5	6	7	8	9	10	11
12	13	14	15	16	17	18
19	20	21	22	23	24	25
26	27	28	29	30	31	

a   If today is the 4th, what will be the day and date in 2 weeks?

b   What day is 9 days after the 13th of May?

c   Which days are there 5 of in the month?

d   If you went on holidays on the 3rd of May for 11 days, on which day would you get back?

e   How many days is it from the 17th to the 23rd of May?

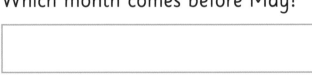

f   Which month comes after May?

Does May always start on a Wednesday?

g   Which month comes before May?

Extended practice

 1

Month:						
Sun	Mon	Tues	Wed	Thur	Fri	Sat

a Fill in the name of the current month.

b Fill in the dates on the correct days.

c What day does the month start on?

d How many days are in the month?

---

2 This calendar shows one month of the year.

Sun	Mon	Tues	Wed	Thur	Fri	Sat
		1	2	3	4	5
6	7	8	9	10	11	12
13	14	15	16	17	18	19
20	21	22	23	24	25	26
27	28	29	30			

a Could it be February?

b Which months could it be?

c How many full weeks are there?

d What date is the third Thursday of the month?

OXFORD UNIVERSITY PRESS

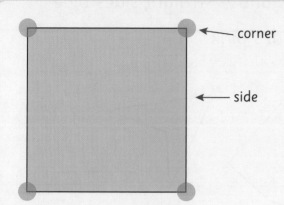

corner

side

The shape has 4 corners and 4 sides.

The sides are straight lines.

How else could you describe this shape?

Guided practice

1 A hexagon has:

a ☐ corners

b ☐ sides

2 A pentagon has:

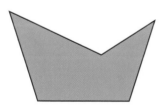

a ☐ corners

b ☐ sides

3 An octagon has:

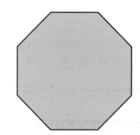

a ☐ corners

b ☐ sides

1 **a** Colour the shapes with 4 corners and 4 straight sides blue.

     b Colour the shapes with curved sides pink.

2 Match the shapes to their names and descriptions.

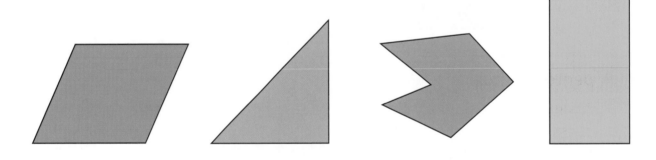

rectangle	hexagon	rhombus	triangle
6 sides and 6 corners	4 sides all the same length	4 corners, opposite sides are the same length	3 straight sides

OXFORD UNIVERSITY PRESS

3 Draw a shape with:

a 3 sides and 3 corners.

b no corners.

c at least 2 straight sides and 1 curved side.

d 5 corners and 5 sides.

e 4 straight sides with 2 sides the same length.

f 4 straight sides of different lengths.

Extended practice

1 Name each shape.

a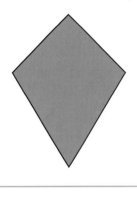

[ ]

b

[ ]

c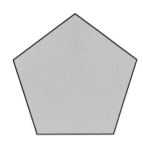

[ ]

d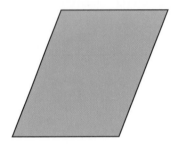

[ ]

e

[ ]

f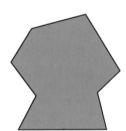

[ ]

2 Which of the shapes from question 1 have ...

4 corners?	5 corners?	8 corners?	no corners?

OXFORD UNIVERSITY PRESS

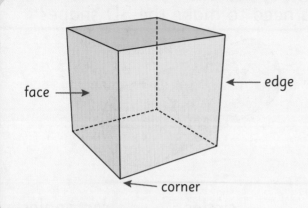

A cube has:

- 6 faces
- 12 edges
- 8 corners.

Faces of 3D shapes can be different 2D shapes, such as circles, triangles or squares.

Guided practice

1

A rectangular prism has:

a [ ] faces

b [ ] edges

c [ ] corners.

2

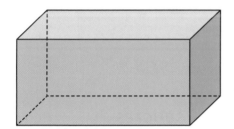

A cylinder has:

a [ ] faces

b [ ] edges

c [ ] corners.

3

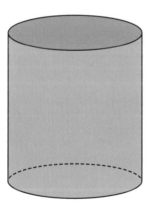

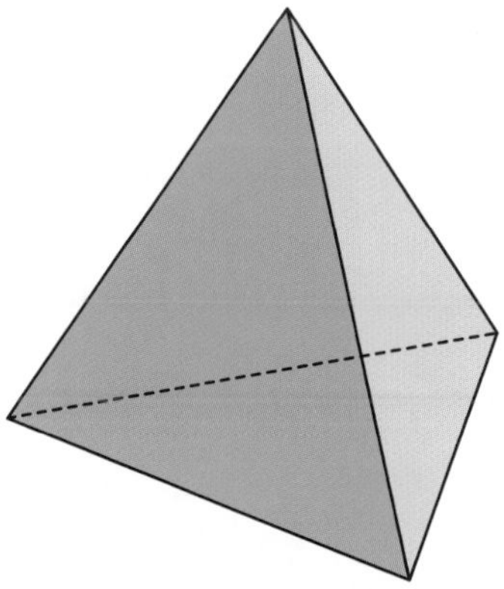

A triangular pyramid has:

a [ ] faces

b [ ] edges

c [ ] corners.

1 How many of each 2D shape do you need to make the 3D shape?

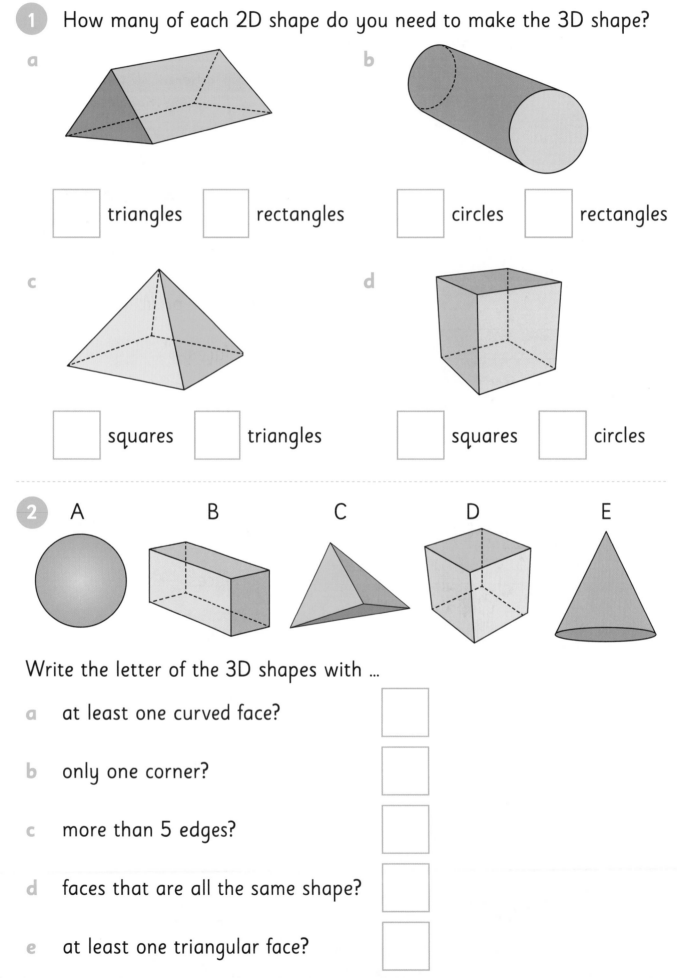

a

☐ triangles   ☐ rectangles

b

☐ circles   ☐ rectangles

c

☐ squares   ☐ triangles

d

☐ squares   ☐ circles

2   A   B   C   D   E

Write the letter of the 3D shapes with ...

a   at least one curved face?   ☐

b   only one corner?   ☐

c   more than 5 edges?   ☐

d   faces that are all the same shape?   ☐

e   at least one triangular face?   ☐

OXFORD UNIVERSITY PRESS

3 Draw lines to match the 3D shapes with their names and descriptions.

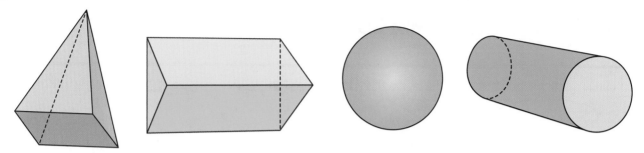

sphere   square pyramid   cylinder   triangular prism

6 corners and 9 edges   1 curved face   3 faces and 2 edges   5 corners, 1 square face and 4 triangular faces

A prism has two parallel bases that are the same shape, and the other faces are rectangles.

4 a Colour the cubes blue.

b Colour the other prisms green.

c Colour the pyramids red.

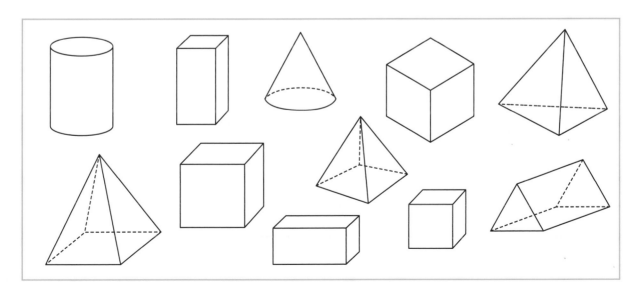

1 Join the dots to make 3D shapes.

a　　　　　　**b**　　　　　　**c**　　　　　　**d**

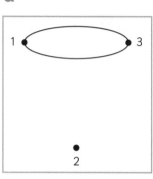 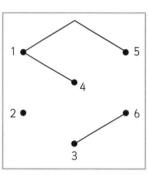 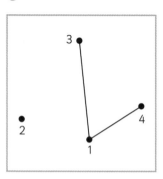 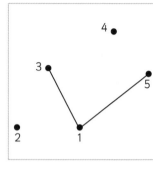

2 Name each 3D shape from question 1.

3 Who am I?

a My faces are rectangles.

I have 8 corners.

I am not a cube.

I am a [　　　　　　].

Draw

b I have 2 edges but no corners.

I have 3 faces.

I am a [　　　　　　].

Draw

OXFORD UNIVERSITY PRESS

The dog is to the right of the clock.

The photo is on the middle shelf.

The train is below the dictionary.

The clock is above the photo and between the dog and the dinosaur.

What other words can you use to describe where something is?

Guided practice

1

a What is above the picnic?

b What is between the slide and the bin?

c Where is the picnic basket?

d What is to the right of the dog?

e What is on the slide?

1 Where is ...

a the computer?

b the whiteboard?

c the teacher?

d the water bottle?

2

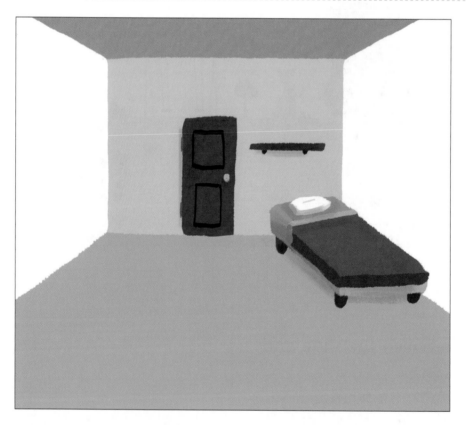

a Draw a clock on the shelf.

b Draw a mat in front of the door.

c Draw a chair next to the bed.

d Draw a desk in the top left corner.

e Draw a bookcase in the bottom right corner.

f Draw a TV to the left of the bookcase.

OXFORD UNIVERSITY PRESS

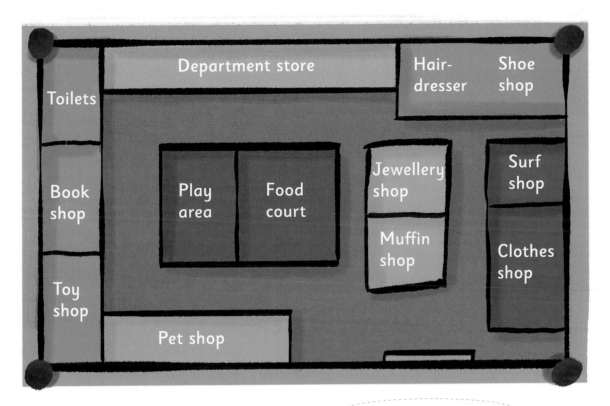

3 Fill in the gaps.

Why do shopping centres have maps?

a The shoe shop is [_____] the hairdresser.

b The book shop is [_____] the toy shop and the toilets.

c The food court is [_____] the department store.

d The play area is [_____] the food court.

4

a What would you go past to get from the pet shop to the department store?

[_____]

b Which way would you turn to get from the surf shop to the muffin shop? [_____]

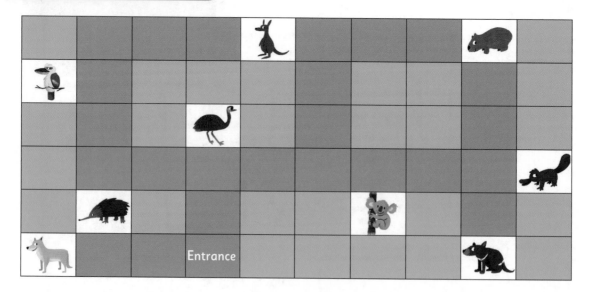

Entrance

1   Where are you?

a   Start at the dingo. Travel 3 squares to the right. Turn left and travel 2 squares.

b   Start at the entrance. Walk 2 squares straight ahead. Turn right. Walk 3 more squares.

2   Write directions to walk along the path from:

a   the entrance to the platypus.

b   the koala to the kookaburra.

c   the echidna to the Tasmanian devil.

OXFORD UNIVERSITY PRESS

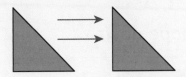

slide

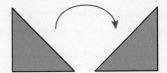

horizontal flip

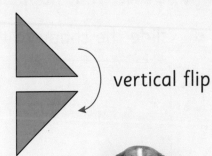

vertical flip

Guided practice

How does a slide change a shape? What about a flip?

1 Slide or flip?

a

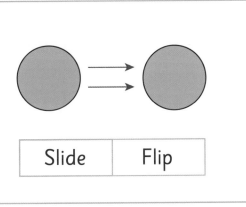

Slide	Flip

b

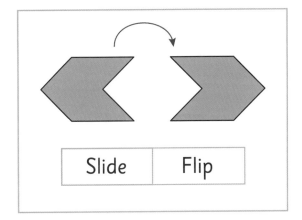

Slide	Flip

c

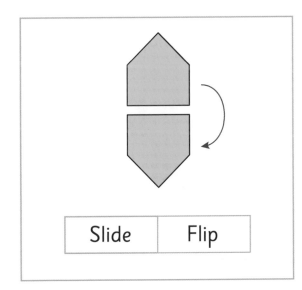

Slide	Flip

d

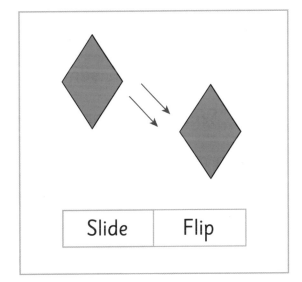

Slide	Flip

1 Draw what happens if you …

a slide the shape to the right.

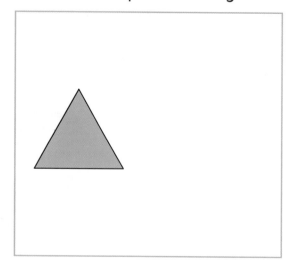

b slide the shape down.

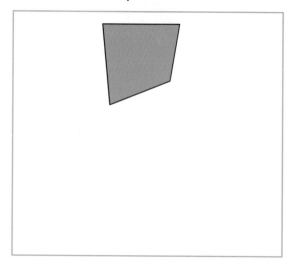

c flip the shape horizontally.

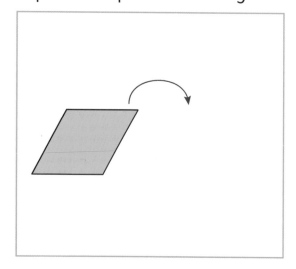

d flip the shape vertically.

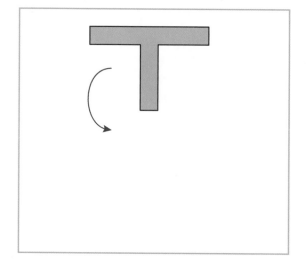

e slide the shape to the left.

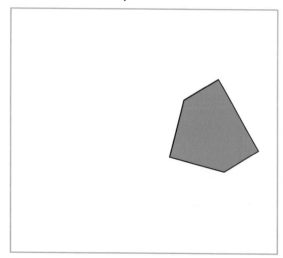

f flip the shape horizontally.

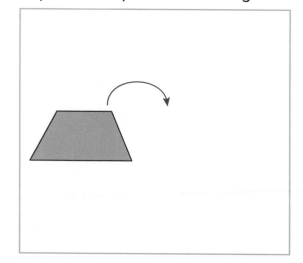

OXFORD UNIVERSITY PRESS

2 Write *slide*, *flip* or *either*.

> Some shapes look the same when they are flipped.

a

R R

b

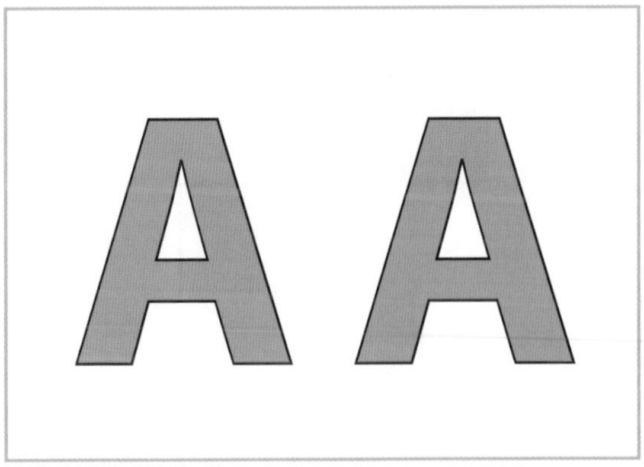

A A

c

B ꓭ

d

F F

e

H H

f

M W

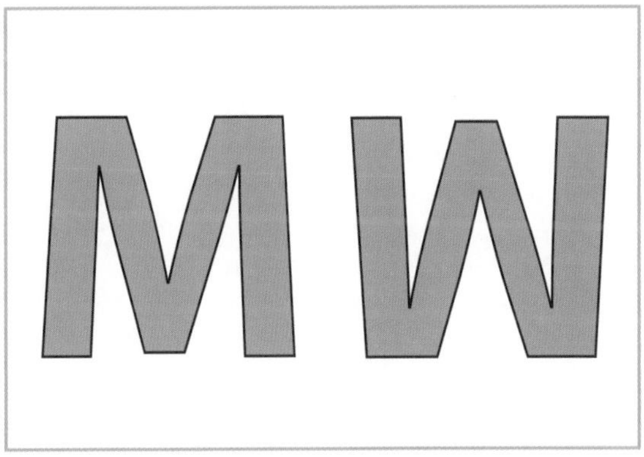

Extended practice

1 Slide then flip each shape.

a

Slide | Flip

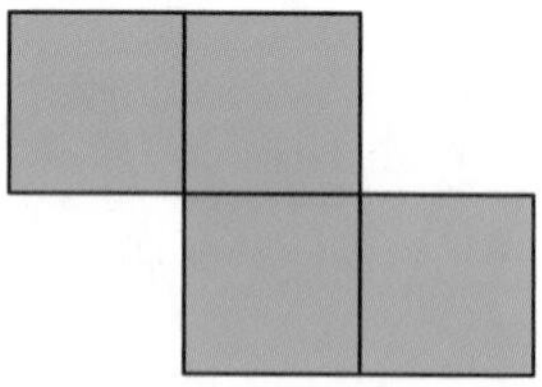

b

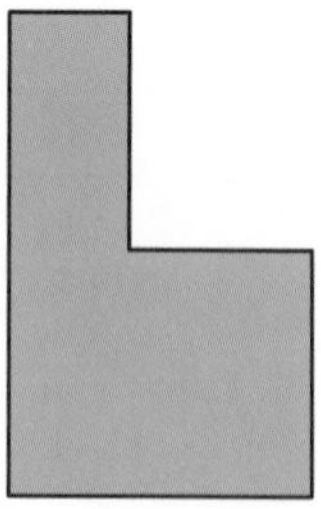

c

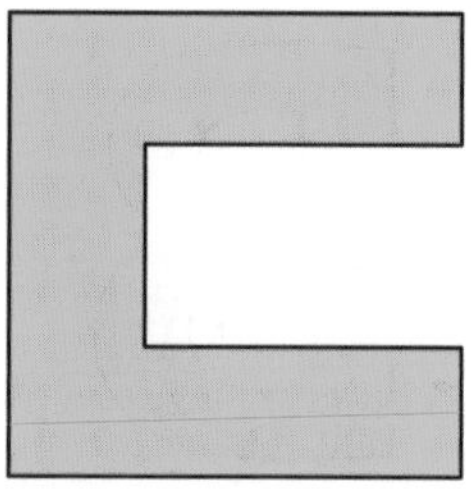

d

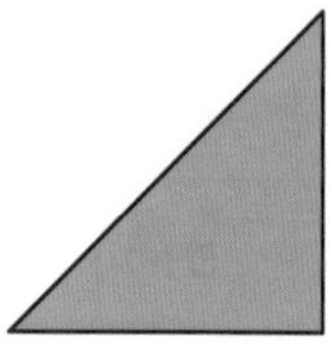

e

OXFORD UNIVERSITY PRESS

half turn

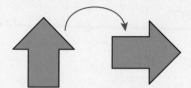

quarter turn
to the right

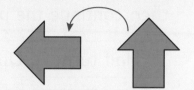

quarter turn
to the left

Guided practice

When we turn something around to the right, this is also known as clockwise as it's the same direction that the hands go around a clock.

1 Half turn or quarter turn?

a

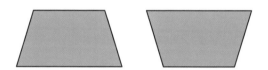

Half turn	Quarter turn

b

Half turn	Quarter turn

c

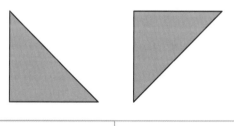

Half turn	Quarter turn

d

Half turn	Quarter turn

e

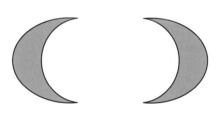

Half turn	Quarter turn

f

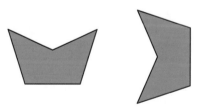

Half turn	Quarter turn

1 Decide whether the pattern is showing half turns or quarter turns, then continue the pattern.

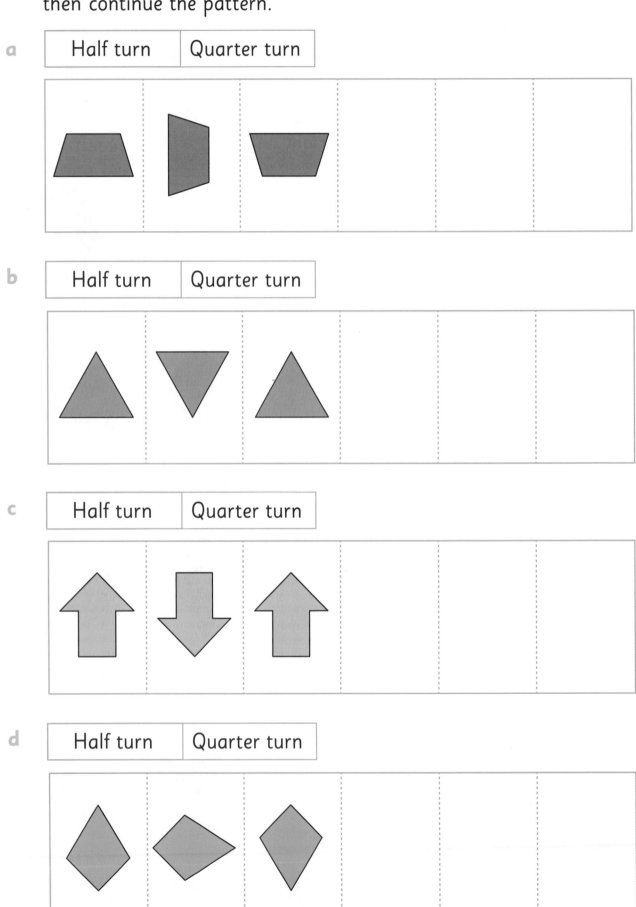

a

| Half turn | Quarter turn |

b

| Half turn | Quarter turn |

c

| Half turn | Quarter turn |

d

| Half turn | Quarter turn |

OXFORD UNIVERSITY PRESS

2 Draw what happens if you do a ...

a half turn.

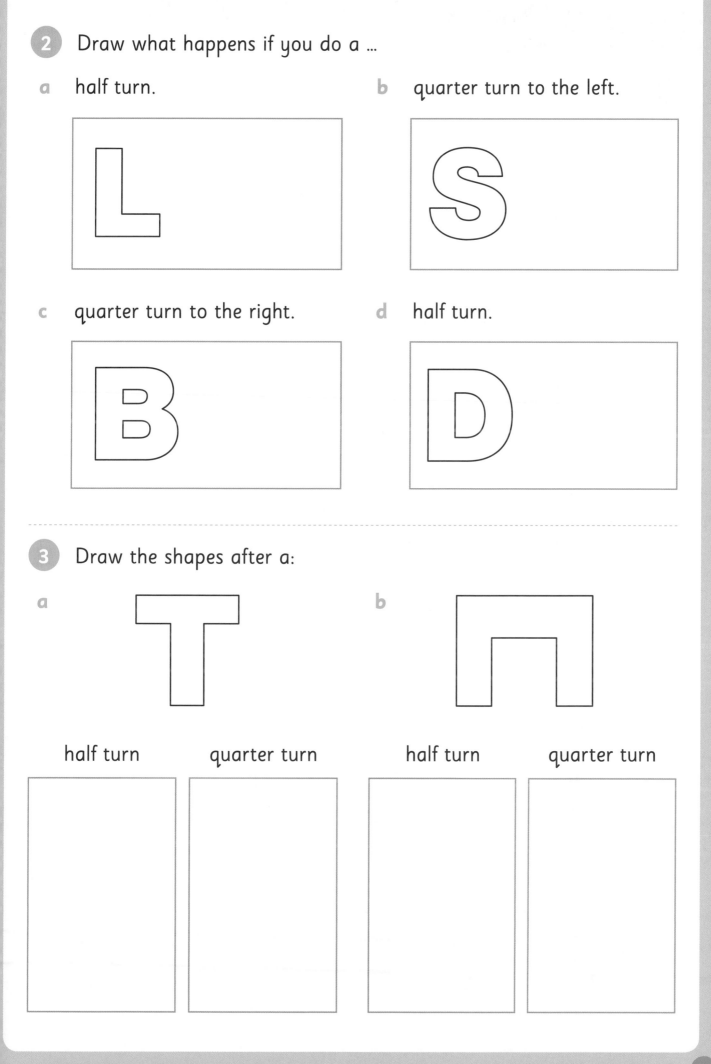

b quarter turn to the left.

c quarter turn to the right.

d half turn.

- -

3 Draw the shapes after a:

a

b

half turn   quarter turn

half turn   quarter turn

1 **a** Describe the turn used to make the pattern.

b Continue the pattern.

2 **a** Describe the turns used to make the pattern.

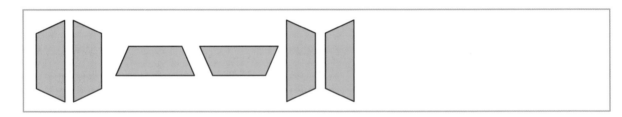

b Continue the pattern.

3 Circle the shapes that show a quarter turn.

a

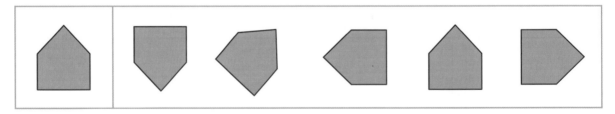

b

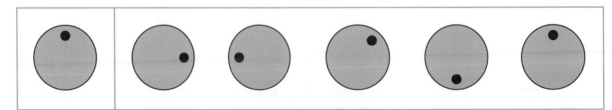

OXFORD UNIVERSITY PRESS

What did you have for dinner?

Dinner	Students
Chicken	Caleb, Serena, Miles
Pizza	Ava, Zac, Josh, Emily, Tayla, Hannah, Sophia, Joseph, Jessie, Caitlin, Casey
Pasta	Riley, Ethan, Toni, Kyle, Matt, Demi, Mason, Darlean

Dinner	Number of students	Total
Chicken	✓✓✓	3
Pizza	✓✓✓✓ ✓✓✓✓ ✓✓✓	11
Pasta	✓✓✓✓ ✓✓✓✓	8

Guided practice

1 Use the information below to complete the table on page 118.

Ice-cream flavour	Students
Vanilla	James, Brittany, Rhys, Georgia, Natalie, Erica, Marco, Ramiz, Olivia, Alicia, Jesse, Mia, Chris, Katie, Emett, Tony
Chocolate	Dylan, Sam, Zoe, Claudia, William, Mason, Violet, Jensen, Laney, Emily, Riley, Felix, Ben, Emma, Matt, Imogen, Steph, Rachael
Mint	Andrew, Kyle, Alex, Penny, Jack, Brenton, Jarrod, Amy, Nathan, Rachael, Casey, McKenzie
Strawberry	Amber, Scarlet, Joey, Kristian, Luke, Bryce, Hannah, Grace

Ice-cream flavour	Number of student (ticks)	Total (number)
Vanilla		
Chocolate		
Mint		
Strawberry		

2 Write a question to match the data.

Independent practice

1

Do you have a brother?
no, no, yes, no, yes, yes, yes, no, no, yes, no, no, yes, yes, no, yes, yes, yes

Count and record

Do you have a brother?	
Yes	No

2 What pet do you have?

Cat	✓✓✓✓✓✓
Dog	✓✓✓✓✓✓✓✓✓✓
Reptile	✓✓
Other	✓✓✓✓✓
None	✓✓✓

a Count and record.

What pet do you have?	
Cat	
Dog	
Reptile	
Other	
None	

b How many students were asked? ☐☐

OXFORD UNIVERSITY PRESS

3 Collect data from 12 students in your class.

Do you have a sister?

Yes	No

Which way is easier to record the data? Why?

4 Record the favourite sport of 12 people in your class.

Sport	Students
Football	
Rugby	
Netball	
Cricket	
Basketball	
Other	

5 What question did you ask to get the data in question 4?

6 Which sport was the:

a most popular? _____   b least popular? _____

1 Write a yes/no question to ask your classmates.

2 Ask 12 people and record their answers.

Name	Results

3 Record the results another way.

Yes	No

Were the results what you expected? Why or why not?

OXFORD UNIVERSITY PRESS

Fruit	Vegetables
卌 卌 ‖	卌 ‖
12	7

Why do some of the tally marks have a diagonal line across them?

Guided practice

1. Count the tally marks.

Favourite drinks			
Milk	**Water**	**Orange juice**	**Soft drink**
‖‖	卌 ‖‖	卌 卌 卌	卌 卌 卌 ‖
Total			

2. Use tally marks to record the colours.

Red	Blue	Green
Total		

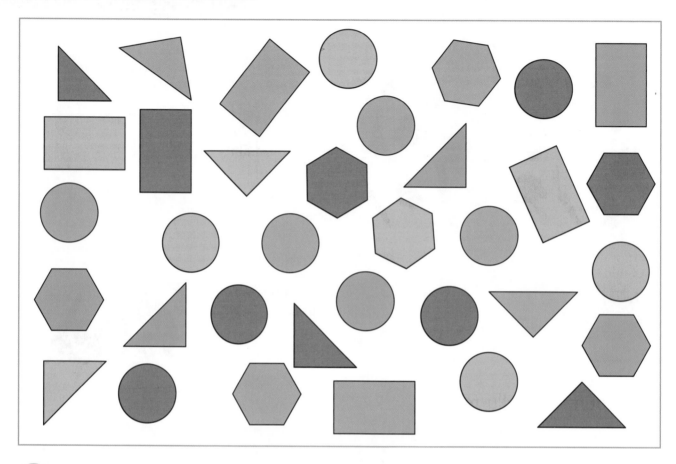

1 **a** Choose a way to sort the shapes into 4 different groups. Record your categories in the table.

Categories:			
Total			

b Use tally marks to count the items in each of your categories.

c Total your tally marks.

d Which category had the most items in it?

OXFORD UNIVERSITY PRESS

2

a Choose 4 sports that are popular in your class and record them in the table.

Sports:				
Tally				
Totals				

b Survey at least 10 people in your class and keep a tally of their answers.

Don't forget to group your tally marks in fives to make them easy to count.

c Total the tallies.

3 Answer these questions about your results.

a Which sport was the most popular?

b Which sport was the least popular?

c What other sports could you have included?

1

a List 3 different ways you could sort the animals.

b Choose one way to sort the data, and then create a table and make a tally for each category.

c Record the total for each category.

OXFORD UNIVERSITY PRESS

What did you do on Saturday afternoon?

7		🏀		
6		🏀		
5		🏀	🎒	
4		🏀	🎒	
3	🍿	🏀	🎒	
2	🍿	🏀	🎒	📕
1	🍿	🏀	🎒	📕
	Movies	**Sport**	**Shopping**	**Reading**

Three people went to the movies.

Seven people played sport.

Three more people went shopping than read.

Sport was the most popular activity.

Seventeen people were surveyed.

Guided practice

How can you tell how many people were surveyed altogether?

1 Answer the questions about the graph.

Bugs in the school garden

10				🐜
9				🐜
8	🐌			🐜
7	🐌			🐜
6	🐌			🐜
5	🐌			🐜
4	🐌	🐛		🐜
3	🐌	🐛		🐜
2	🐌	🐛	🪱	🐜
1	🐌	🐛	🪱	🐜
	Snails	**Slugs**	**Worms**	**Ants**

a   Which bug was there the most of?

b   The least?

c   How many more snails than slugs were found?

d   How many bugs were found in total?

1

a Use the data in the table to complete the pictograph.

5						
4						
3						
2	▬					
1	▬					
	Tim	Devon	Mai	Rex	Tina	Poh

Number of hours watching TV last night					
Tim	**Devon**	**Mai**	**Rex**	**Tina**	**Poh**
2	4	0	2	5	1

b Which 2 people watched the same amount of TV last night?

c Who watched the most?

d Who watched the least?

2 Hair colour in a Year 2 class

Hair colour	Tally	Total
Brown	卌 I	
Blond	IIII	
Black	卌 III	
Red	II	

a Record the totals in the table.

b Finish the pictograph.

8	
7	
6	☺
5	☺
4	☺
3	☺
2	☺
1	☺
	Brown

OXFORD UNIVERSITY PRESS

3

a Ask 10 students in your class if they take swimming lessons.
Record the results in a list (yes/no).

b Make a tally table using the results.

Answer	Tally	Total

Which method of displaying the data do you find the easiest to understand?

c Use ticks to show the results.

	1	2	3	4	5	6	7	8	9	10
Yes										
No										

d Write one statement about the results.

Extended practice

a Make a table using the data in the graph.

How I get to school

10				
9	Car			
8	Car			
7	Car		Walk	
6	Car		Walk	
5	Car	Bike	Walk	
4	Car	Bike	Walk	
3	Car	Bike	Walk	
2	Car	Bike	Walk	
1	Car	Bike	Walk	Bus
	Car	**Bike**	**Walk**	**Bus**

Transport	Tally	Total

b How many people ride their bikes?

c Do more people walk or catch the bus?

d How many more people walk than ride bikes?

e Write a question of your own about the data.

OXFORD UNIVERSITY PRESS

An elephant will fly a helicopter over the school.

A pop star will visit today.

You will eat lunch today.

You will do mathematics today.

| Impossible | Less likely | Most likely | Certain |

Guided practice

What does "certain" mean?

1 Most likely or less likely today?

a sport lesson

| Most likely | Less likely |

b dance lesson

| Most likely | Less likely |

2 Certain or impossible today?

a wearing shoes

| Certain | Impossible |

b the school day will end

| Certain | Impossible |

1 Match the words to the situations.

Impossible	Less likely	Most likely	Certain

a It will rain today.

b It will be cold today.

c It will get dark tonight.

d You will go shopping today.

e You will have pizza for dinner.

f There is a live dinosaur in the playground.

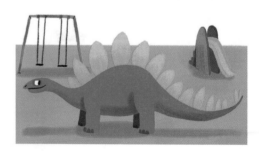

OXFORD UNIVERSITY PRESS

2 Draw a line to show how likely each event is.

Impossible   Less likely   Most likely   Certain

a

You will drink water today.

b

You will have a birthday this year.

c

You will travel in a plane today.

d

You will drive a train today.

Would everyone be likely to drink water today?

3 Suggest an event that is:

a impossible.

b less likely.

c most likely.

d certain.

1 Describe the chances of pulling out a:

a red teddy.

b blue teddy.

c yellow teddy.

---

2 Match the descriptions to the boxes.

a certain to pick a red ball

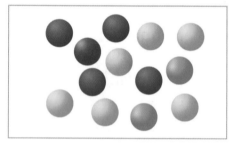

b impossible to pick a red ball

c less likely to pick a green ball

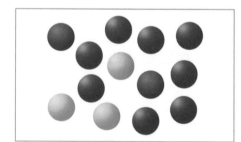

d most likely to pick a blue ball

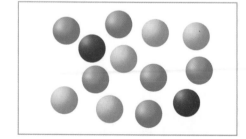

OXFORD UNIVERSITY PRESS

GLOSSARY

addition The joining or adding of two numbers together to find the total. Also known as *adding*, *plus* and *sum*.

Example:

★★★ + ★★ = ★★★★★

3 and 2 is 5

anticlockwise Moving in the opposite direction to the hands on a clock.

area The size of an object's surface.

Example:
It takes 12 tiles to cover this placemat.

array An arrangement of items into even columns and rows that make them easier to count.

balance scale Equipment that balances items of equal mass – used to compare the mass of different items. Also called pan balance or equal arm balance.

base The bottom edge of a 2D shape or the bottom face of a 3D shape.

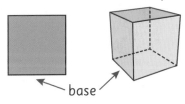

base

calendar A chart or table showing the days, dates, weeks and months in a year.

Month —— **January 2017** —— Year

Sun	Mon	Tues	Wed	Thur	Fri	Sat
1	2	3	4	5	6	7
8	9	10	11	12	13	14
15	16	17	18	19	20	21
22	23	24	25	26	27	28
29	30	31				

Day —— Sun... Date —— 15

capacity The amount that a container can hold.

Example:
The jug has a capacity of 4 cups.

4 cups
3 cups
2 cups
1 cup

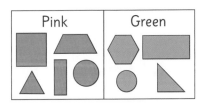

category A group of people or things sharing the same characteristics.

Pink	Green

centimetre A unit for measuring the length of smaller items.

Example: Length is 15 cm.

80 cm

circle   A 2D shape with a continuous curved line that is always the same distance from the centre point.

clockwise   Moving in the same direction as the hands on a clock.

cone   A 3D shape with a circular base that tapers to a point.

corner   The point where two edges of a shape or object meet.

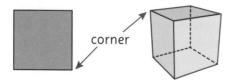

corner

cube   A rectangular prism where all 8 faces are squares of equal size.

cylinder   A 3D shape with 2 parallel circular bases and one curved surface.

data   Information gathered through methods such as questioning, surveys or observation.

day   A period of time that lasts 24 hours.

difference (between)   A form of subtraction or take away.

Example: The difference between 11 and 8 is 3.

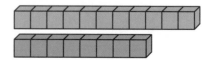

digit   The single numerals from 0 to 9. They can be combined to make larger numbers.

Example: 24 is a 2-digit number.

378 is a 3-digit number.

division/dividing   Sharing into equal groups.

Example: 9 divided by 3 is 3

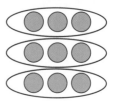

double/doubles   Adding two identical numbers or multiplying a number by 2.

Example: 4 + 4 = 8   2 x 4 = 8

OXFORD UNIVERSITY PRESS

duration   How long something lasts.

Example: The school week lasts for 5 days.

edge   The side of a shape or the line where two faces of an object meet.

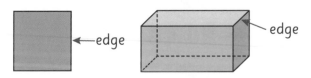

eighth   One part of a whole or group divided into eight equal parts.

Eighth of a whole   Eighth of a group

equal   Having the same number or value.

Example:

Equal size   Equal numbers

equation   A written mathematical problem where both sides are equal.

Example: 4 + 5 = 6 + 3

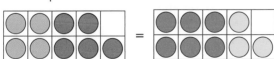

estimate   A thinking guess.

face   The flat surface of a 3D shape.

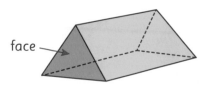

face

flip   To turn a shape over horizontally or vertically. Also known as reflection.

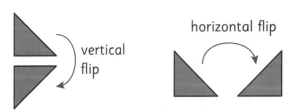

vertical flip

horizontal flip

fraction   An equal part of a whole or group.

Example: One out of two parts or $\frac{1}{2}$ is shaded.

friendly numbers   Numbers that are easier to add to or subtract from.

Example: 10, 20 or 100

half   One part of a whole or group divided into two equal parts. Also used in time for 30 minutes.

Example:

Half of   Half of   Half past 4
a whole   a group

hexagon   A 2D shape with 6 sides.

horizontal   Parallel with the horizon or going straight across.

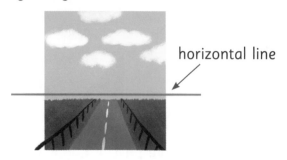

horizontal line

jump strategy   A way to solve number problems that uses place value to "jump" along a number line by hundreds, tens and ones.

Example: 16 + 22 = 38

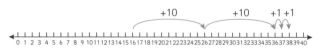

length   How long an object is from end to end.

Example: This poster is 3 pens long.

mass   How heavy an object is.

heavy   light

metre   A unit for measuring the length of larger objects.

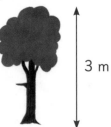

3 m

month   The time it takes the moon to orbit the Earth. There are 12 months in a year.

January   February   March
April   May   June
July   August   September
October   November   December

near doubles   A way to add two nearly identical numbers by using known doubles facts.

Example: 4 + 5 = 4 + 4 + 1 = 9

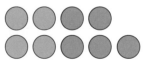

number line   A line on which numbers can be placed to show their order in our number system or to help with calculations.

0   10   20   30   40   50   60   70   80   90   100

number sentence   A way to record calculations using numbers and mathematical symbols.

Example: 23 + 7 = 30

numeral   A figure or symbol used to represent a number.

Example:

1 – one   2 – two   3 – three

OXFORD UNIVERSITY PRESS

octagon　A 2D shape with 8 sides.

ordinal numbers　Numbers that show the order or position of something in relation to others.

1st　2nd　3rd　4th　5th　6th

pair　Two items that go together.

Example: Pairs that make 4

2 and 2　　　3 and 1

Pair of socks

parallel lines　Straight lines that are the same distance apart and so will never cross.

parallel　　　parallel　　　not parallel

partitioning　Dividing or separating an amount into parts.

Example: Some of the ways 10 can be partitioned are:

5 and 5　　　4 and 6　　　9 and 1

pattern　A repeating design or sequence of numbers.

Example: Shape pattern

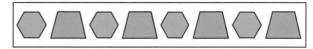

Number pattern

2, 4, 6, 8, 10, 12

pentagon　A 2D shape with 5 sides.

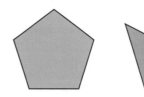

pictograph　A way of representing data using pictures to make it easy to understand.

Example: Favourite juices in our class

place value　The value of a digit depending on its place in a number.

Hundreds	Tens	Ones
		8
	8	6
8	6	3

position　Where something is in relation to other items.

Example: The boy is under the tree that is next to the house.

prism   A 3D shape with parallel bases of the same shape and rectangular side faces.

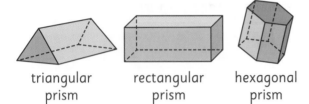

triangular prism   rectangular prism   hexagonal prism

pyramid   A 3D shape with a 2D shape as a base and triangular faces meeting at a point.

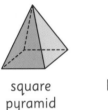

square pyramid   hexagonal pyramid

quadrilateral   Any 2D shape with four sides.

quarter   One part of a whole or group divided into four equal parts. Also used in time for 15 minutes.

Example:

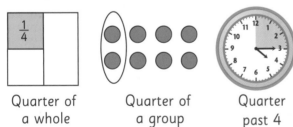

Quarter of a whole   Quarter of a group   Quarter past 4

rectangle   A 2D shape with four sides and four right angles. The opposite sides are parallel and equal in length.

right angle

rhombus   A 2D shape with four sides, all of the same length and opposite sides parallel.

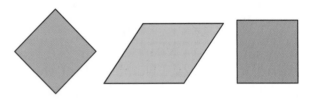

skip counting   Counting forwards or backwards by the same number each time.

Example: Skip counting by 5s: 5, 10, 15, 20, 25, 30

Skip counting by 2s: 1, 3, 5, 7, 9, 11, 13

slide   To move a shape to a new position without flipping or turning it. Also known as *translate*.

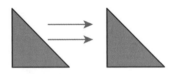

sphere   A 3D shape that is perfectly round.

split strategy   A way to solve number problems that involves splitting numbers up using place value to make them easier to work with.

Example: 21 + 14 = 35

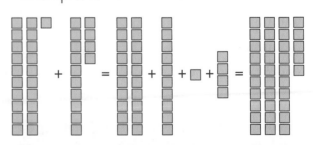

OXFORD UNIVERSITY PRESS

square A 2D shape with four sides of equal length and four right angles. A square is a type of rectangle.

right angle

strategy A way to solve a problem. In mathematics you can often use more than one strategy to get the right answer.

Example: 32 + 27 = 59

Jump strategy

Split strategy

30 + 2 + 20 + 7 = 30 + 20 + 2 + 7 = 59

subtraction The taking away of one number from another number. Also known as *subtracting*, *take away*, *difference between* and *minus*.

Example: 5 take away 2 is 3

survey A way of collecting data or information by asking questions.

Strongly agree	☐
Agree	☑
Disagree	☐
Strongly disagree	☐

table A way to organise information that uses columns and rows.

Flavour	Number of people
Chocolate	12
Vanilla	7
Strawberry	8

tally marks A way of keeping count that uses single lines with every fifth line crossed to make a group.

three-dimensional or 3D A shape that has three dimensions – length, width and depth. 3D shapes are not flat.

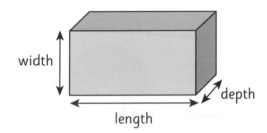
width
depth
length

trapezium A 2D shape with four sides and only one set of parallel lines.

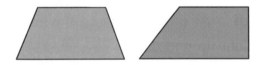

triangle A 2D shape with three sides.

turn Rotate around a point.

two-dimensional or 2D A flat shape that has two dimensions – length and width.

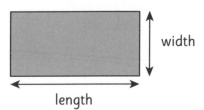
width
length

unequal   Not having the same size or value.

Example:

   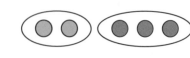

Unequal size   Unequal numbers

value   How much something is worth.

Example:

This coin is   This coin is
worth 5c.   worth $1.

vertical   At a right angle to the horizon or straight up and down.

vertical line

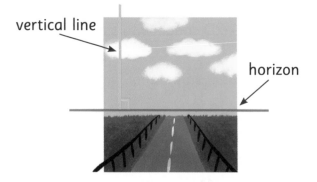

horizon

volume   How much space an object takes up.

Example: This object has a volume of 4 cubes.

week   A period of time that lasts 7 days.

Monday   Tuesday   Wednesday

Thursday   Saturday   Sunday
           Friday

whole   All of an item or group.

Example:

A whole shape   A whole group

width   How wide an object is from one side to the other.

Example: This poster is 2 pens wide.

year   The time it takes the Earth to orbit the Sun, which is approximately 365 days.

OXFORD UNIVERSITY PRESS

UNIT 1: Topic 1

Guided practice

1   a 10   b 24   c 100   d 135

2   a 263   b 425   c 617

Independent practice

1   3 hundreds, 5 tens, 4 ones

2   2 hundreds, 0 tens, 6 ones

3   4 hundreds, 2 tens, 3 ones

4   a 4 tens   8 hundreds

    b 7 ones   3 hundreds

    c 4 hundreds   1 ten   3 ones

    d 0 tens   5 hundreds   8 ones

5   Teacher to check. Teacher: Look for answers that show students' ability to correctly interpret and represent hundreds, tens and ones with base 10 materials, an abacus or any other simplified means that doesn't involve drawing each separate one.

Extended practice

1   a 863   b 368   c 638

    d 38, 36, 86, 83, 68, 63

2   Teacher to check. Teacher: Look for answers that show students' ability to manipulate their chosen digits to make the biggest and smallest 3-digit numbers possible.

3   a 11   b 141, 207, 297, 279

UNIT 1: Topic 2

Guided practice

1   a $8 + 9 = 8 + 8 + 1 = 16 + 1 = 17$

    b $10 + 13 = 10 + 10 + 3 = 20 + 3 = 23$

    c $15 + 17 = 15 + 15 + 2 = 30 + 2 = 32$

    d $12 + 13 = 12 + 12 + 1 = 24 + 1 = 25$

    e $14 + 15 = 14 + 14 + 1 = 28 + 1 = 29$

Independent practice

1   a $7 + 7 = 14$ b $11 + 11 = 22$

    c $16 + 16 = 32$ d $20 + 20 = 40$

    e $25 + 25 = 50$ f $50 + 50 = 100$

2   a $7 + 9 = 7 + 7 + 2 = 14 + 2 = 16$

    b $11 + 13 = 11 + 11 + 2 = 22 + 2 = 24$

---

    c $16 + 17 = 16 + 16 + 1 = 32 + 1 = 33$

    d $20 + 23 = 20 + 20 + 3 = 40 + 3 = 43$

    e $25 + 27 = 25 + 25 + 2 = 50 + 2 = 52$

Guided practice

1   a $17 + 6 = 17 + 3 + 3$
             $= 20 + 3$
             $= 23$

    b $29 + 8 = 29 + 1 + 7$
             $= 30 + 7$
             $= 37$

    c $38 + 5 = 38 + 2 + 3$
             $= 40 + 3$
             $= 43$

Independent practice

1   a

    $18 + 9 = 18 + 2 + 7$
             $= 20 + 7$
             $= 27$

    b

    $26 + 7 = 26 + 4 + 3$
             $= 30 + 3$
             $= 33$

2   a $25 + 8 = 25 + 5 + 3$
             $= 30 + 3$
             $= 33$

    b $37 + 6 = 37 + 3 + 3$
             $= 40 + 3$
             $= 43$

Extended practice

1   a 42   b 53   c 63   d 82

    e 101   f 112

2   a 37   b 44   c 52   d 66

    e 72   f 85

3   a 46   Getting to a 10

    b 81   Near doubles

    c 202   Near doubles

UNIT 1: Topic 3

Guided practice

1
$9 + 5 = 14$

2
$11 + 7 = 18$

3
$6 + 18 = 24$

4
$23 + 5 = 28$

Independent practice

1
$19 + 8 = 27$

2
$24 + 6 = 30$

3
$7 + 14 = 21$

4
$5 + 21 = 26$

5
$32 + 10 = 42$

Extended practice

1   Teacher to check. Teacher: Look for answers that show students' ability to accurately space their numbers and correctly represent the addition sum.

    a $14 + 5 = 19$ b $21 + 6 = 27$

    c $32 + 7 = 39$

2   Teacher to check. Teacher: Look for answers that show students' ability to understand that they can partition numbers into 10s to add more easily or who use skip counting in their jumps rather than making steps of 1.

    a $23 + 12 = 35$ b $35 + 24 = 59$

Guided practice

1 $5 + 4 = 4 + 5 = 9$

2 $7 + 2 + 4 = 7 + 4 + 2 = 13$

3 $4 + 6 + 5 = 4 + 5 + 6$ OR $6 + 5 + 4 = 15$

Independent practice

1 Teacher to check. Teacher: Look for answers that show students' ability to correctly represent each part of the equation and who can move the numbers around to make new equations.

 a Possible combinations: $7 + 1 + 5$, $5 + 1 + 7$, $5 + 7 + 1$, $1 + 5 + 7$, $1 + 7 + 5 = 13$

 b Possible combinations: $2 + 4 + 9$, $4 + 9 + 2$, $4 + 2 + 9$, $9 + 4 + 2$, $9 + 2 + 4 = 15$

 c Possible combinations: $8 + 7 + 1$, $7 + 8 + 1$, $7 + 1 + 8$, $1 + 8 + 7$, $1 + 7 + 8 = 16$

2 a Possible combinations: $5 + 3 + 6$, $6 + 5 + 3$, $6 + 3 + 5$, $3 + 5 + 6$, $3 + 6 + 5 = 14$

 b Possible combinations: $4 + 5 + 7$, $5 + 7 + 4$, $5 + 4 + 7$, $7 + 4 + 5$, $7 + 5 + 4 = 16$

 c Possible combinations: $1 + 4 + 9$, $4 + 9 + 1$, $4 + 1 + 9$, $9 + 4 + 1$, $9 + 1 + 4 = 14$

 d Possible combinations: $7 + 9 + 8$, $9 + 8 + 7$, $9 + 7 + 8$, $8 + 9 + 7$, $8 + 7 + 9 = 24$

Extended practice

1 a $8 + 5 + 7 = 5 + 8 + 7 = 20$

 b $6 + 9 + 4 = 4 + 9 + 6 = 19$

 c $8 + 3 + 4 = 4 + 3 + 8$ OR $4 + 8 + 3 = 15$

 d $22 = 9 + 7 + 6 = 6 + 7 + 9$

 e $19 = 8 + 4 + 7 = 7 + 8 + 4$

2 Teacher to check. Teacher: Look for answers that show students' ability to make three different combinations that add up to the correct total.

UNIT 1: Topic 4

Guided practice

Subtract by getting to a 10

1 a $13 - 4 = 13 - 3 - 1$
         $= 10 - 1$
         $= 9$

 b $21 - 5 = 21 - 1 - 4$
         $= 20 - 4$
         $= 16$

 c $32 - 5 = 32 - 2 - 3$
         $= 30 - 3$
         $= 27$

Independent practice

1 a

$14 - 8 = 14 - 4 - 4$
        $= 10 - 4$
        $= 6$

 b

$25 - 7 = 25 - 5 - 2$
        $= 20 - 2$
        $= 18$

2 a $35 - 8 = 35 - 5 - 3$
         $= 30 - 3$
         $= 27$

 b $34 - 7 = 34 - 4 - 3$
         $= 30 - 3$
         $= 27$

Guided practice

1 a Count up from 5 to 10
      $5 + 5 = 10$

   Count up from 10 to 13
   $10 + 3 = 13$

   The difference between 13 and 5 is $5 + 3$ OR 8.
   So $13 - 5 = 8$

 b Count up from 19 to 20
    $19 + 1 = 20$

   Count up from 20 to 24
   $20 + 4 = 24$

   The difference between 19 and 24 is $1 + 4$ OR 5.
   So $24 - 19 = 5$

Independent practice

1 a $14 - 8$
    Count up from 8 to 10
    $8 + 2 = 10$

   Count up from 10 to 14
   $10 + 4 = 14$

   The difference between 14 and 8 is $2 + 4$ OR 6.
   So $14 - 8 = 6$

 b $23 - 17$
    $17 + 3 = 20$
    $20 + 3 = 23$
    So $23 - 17 = 6$

2 a $16 - 9 = 7$

 b $25 - 19 = 6$

Extended practice

1 a $12 - 4 = 8$ b $15 - 8 = 7$

   c $21 - 9 = 12$ d $32 - 6 = 26$

   e $46 - 7 = 39$ f $53 - 5 = 48$

2 a $18 - 7 = 11$ b $22 - 15 = 7$

   c $35 - 23 = 12$ d $38 - 27 = 11$

   e $43 - 36 = 7$ f $48 - 29 = 19$

3 a $28 - 16 = 12$ Counting up

 b $34 - 8 = 26$ Getting to a 10

 c $41 - 34 = 7$ Counting up

UNIT 1: Topic 5

Guided practice

1 $14 - 6 = 8$

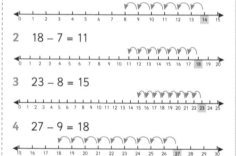

2 $18 - 7 = 11$

3 $23 - 8 = 15$

4 $27 - 9 = 18$

Independent practice

1 $28 - 7 = 21$

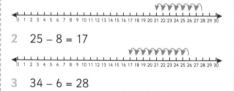

2 $25 - 8 = 17$

3 $34 - 6 = 28$

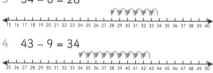

4 $43 - 9 = 34$

5 $48 - 12 = 36$

Extended practice

1 Teachers to check number lines. Teachers: Look for answers that show students' ability to accurately space their numbers and correctly represent

OXFORD UNIVERSITY PRESS

the subtraction sum. Students may also use skip counting or partitioning to show the steps taken to get the answer.

a 19 − 4 = 15　　b 28 − 8 = 20
c 33 − 7 = 26　　d 36 − 14 = 22

2 a 37 − 4 − 5 = 28

b 41 − 7 − 6 = 28

Guided practice

1 a 16 − 10 = 6　　16 − 6 = 10
b (answers can be in any order)
　19 − 4 = 15　　19 − 15 = 4
c 20 − 9 = 11　　20 − 11 = 9
d (answers can be in any order)
　23 − 7 = 16　　23 − 16 = 7

Independent practice

1 a 13 + 5 = 18　　18 − 5 = 13
　5 + 13 = 18　　18 − 13 = 5
b 16 + 8 = 24　　24 − 8 = 16
　8 + 16 = 24　　24 − 16 = 8
c 25 + 7 = 32　　32 − 7 = 25
　7 + 25 = 32　　32 − 25 = 7

2 a 14 − 8 = 6　　OR　14 − 6 = 8
b 26 − 16 = 10　OR　26 − 10 = 16
c 25 − 12 = 13　OR　25 − 13 = 12
d 38 − 11 = 27　OR　38 − 27 = 11

Extended practice

1 a 11 + 7 = 18　　18 − 7 = 11
　7 + 11 = 18　　18 − 11 = 7
b 15 + 19 = 34　　34 − 15 = 19
　19 + 15 = 34　　34 − 19 = 15

2 Teacher to check. Teacher: Look for answers that show students' ability to correctly represent the equation on the number line using single steps, skip counting or partitioning.
a 24 − 5 = 19　　19 + 5 = 24
b 35 − 12 = 23　　23 + 12 = 35
　OR　12 + 23 = 35

UNIT 1: Topic 6

Guided practice

1 3 + 3 + 3 = 3 × 3 = 9

2 5 + 5 + 5 + 5 + 5 = 5 × 5 = 25

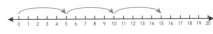

3 3 + 3 + 3 + 3 + 3 = 15
5 threes are 15
5 × 3 = 15

4 5 + 5 + 5 + 5 + 5 + 5 = 30
6 fives are 30
6 × 5 = 30

Independent practice

1 2 × 10 = 20
2 8 × 2 = 16
3 5 × 4 = 20
4 3 × 5 = 15

5 6 × 2 = 12

6 7 × 3 = 21

7 4 × 3 = 12 or 3 × 4 = 12
8 4 × 4 = 16
9 10 × 3 = 30 or 3 × 10 = 30
10 3 × 6 = 18

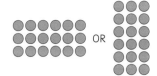
OR

11 4 × 5 = 20

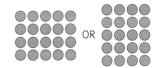
OR

Extended practice

1 14: No　　18: Yes　　20: No
23: No　　21: Yes　　30: Yes

2 20
Teacher to check. Teacher: Look for answers that show ability to use repeated addition or multiplication knowledge to work out the correct answer.

3 Teacher to check. Teacher: Look for answers that show ability to understand that arrays have the same number of items in each row and in each column, and check that arrays match equations.

UNIT 1: Topic 7

Guided practice

1 a 12 divided by 4 is 3　　12 ÷ 4 = 3
b 15 divided by 3 is 5　　15 ÷ 3 = 5

2 a Unequal　　b Equal

Independent practice

1 a

9 − 3 = 6　　6 − 3 = 3　　3 − 3 = 0
9 divided by 3 = 3　　9 ÷ 3 = 3

b

12 − 2 = 10　　10 − 2 = 8　　8 − 2 = 6
6 − 2 = 4　　4 − 2 = 2　　2 − 2 = 0
12 divided by 6 = 2　　12 ÷ 6 = 2

c

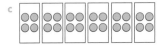

10 − 5 = 5　　5 − 5 = 0
10 divided by 2 = 5　　10 ÷ 2 = 5

2 a

12 ÷ 4 = 3

b

25 ÷ 5 = 5

c

24 ÷ 6 = 4

3 a 9 ÷ 3 = 3
b 20 ÷ 4 = 5

Extended practice

1 Teacher to check. Teacher: Look for answers that show students' ability to match their diagrams to the equations successfully.
The possibilities are 16 ÷ 1 = 16, 16 ÷ 2 = 8, 16 ÷ 4 = 4, 16 ÷ 8 = 2 or 16 ÷ 16 = 1.

2 How many rows of students in a class of:

12? 3   20? 5   28? 7

3

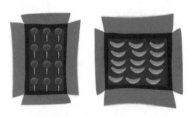

UNIT 1: Topic 8

Guided practice

1   a   8 + 5 = 13 or 5 + 8 = 13

    b   7 + 6 = 13 or 6 + 7 = 13

2   a   15 − 7 = 8

    b   18 − 9 = 9

3   a   7 + 6 = 13

    b   8 + 5 = 13

    c   15 − 7 = 8

    d   5 + 8 = 13

    e   18 − 9 = 9

    f   15 − 8 = 7

Independent practice

1   7 + 7 + 1= 14 + 1 = 15

2   a   9 + 1 + 5 = 10 + 5 = 15

    b   17 − 7 − 1 = 10 − 1 = 9

3 & 4   Teachers may wish to photocopy the tables and have students fill in the addition and subtraction facts that they know first, followed by practice sessions. This will hopefully lead to complete retention of all the necessary addition and subtraction facts.

5   Look for students who use estimation strategies to find the most likely answers.

    a   96

    b   42

    c   74

    d   38

6   Answers may vary. Students could share their ideas with each other. Likely responses are:

    a   Jasmin added instead of subtracting. Correct answer: 8.

    b   Jasmin should have used a strategy of doubling plus 1. Correct answer: 29.

    c   Jasmin took away a 10 as well as 8. Correct answer: 37.

Extended practice

1   a   Tilly

    b   Look for students who use estimation strategies to explain why Tilly has more (almost 100 compared to Billy's 80).

2   Answers will vary.

3   a   360

    b   663 = egg

    c   338 = bee

    d   Answers will vary. Possible responses include:

      638 = beg, 818 = bib, 618 = big, 808 = bob, 608 = bog, 733 = eel, 336 = gee, 771 = ill, 805 = sob.

UNIT 2: Topic 1

Guided practice

1   The whole pizza and the whole apple should have a square drawn around them.

2   The pizza slice and the half apple should have a circle drawn around them.

3

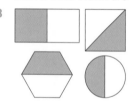

Independent practice

1   a $\frac{1}{4}$ b $\frac{1}{2}$ c $\frac{1}{8}$ d $\frac{1}{4}$

2   a $\frac{1}{8}$ b $\frac{1}{2}$ c $\frac{1}{4}$

3 $\frac{1}{2}$ **4** $\frac{1}{8}$

5   a $\frac{1}{2}$ There are 2 pieces.

    b $\frac{1}{4}$ There are 4 pieces.

    c $\frac{1}{8}$ There are 8 pieces.

6

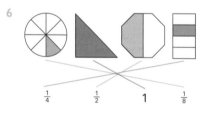

$\frac{1}{4}$ $\frac{1}{2}$ 1 $\frac{1}{8}$

7   a c   b a   c d

Extended practice

1   The following figures should be circled: c and e

2   a–c   Teacher to check: some combination of the following:

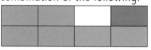

3   Teacher to check. Look for answers that show students' ability to equally divide the shapes and correctly identify the fraction required. Students may draw lines on the shapes to help them find the required fractions. Sample answers:

UNIT 2: Topic 2

Guided practice

1   a   4 items should be coloured in.

    b   5 items should be coloured in.

2   a   1 item should be coloured in.

    b   3 items should be coloured in.

3   a   1 item should be coloured in.

    b   2 items should be coloured in.

Independent practice

1   a $\frac{1}{4}$ b $\frac{1}{2}$ c $\frac{1}{8}$ d $\frac{1}{2}$

2   a $\frac{1}{8}$ b $\frac{1}{2}$ c $\frac{1}{4}$

3   a   2 items should be circled.

      $\frac{1}{2}$ of 4 is 2.

    b   2 items should be circled.

      $\frac{1}{4}$ of 8 is 2.

    c   2 items should be circled.

      $\frac{1}{8}$ of 16 is 2.

    d   3 items should be circled.

      $\frac{1}{4}$ of 12 is 3.

4   a   12 items should be coloured red.

    b   6 items should be coloured blue.

    c   3 items should be coloured green.

Extended practice

1   a $\frac{1}{2}$ of 20   b $\frac{1}{8}$ of 24   c $\frac{1}{2}$ of 16

2   a $\frac{1}{4}$ b $\frac{1}{2}$ c $\frac{1}{8}$

OXFORD UNIVERSITY PRESS

UNIT 3: Topic 1

Guided practice

1

How many of these ...	... do you need to make this?	Draw the answer	Write the answer
			5c + 5c = 10c
			50c + 50c + 50c + 50c = $2 OR 4 × 50c = $2
			20c + 20c + 20c + 20c + 20c = $1 OR 5 × 20c = $1

2

How many of these ...	... do you need to make this?	Draw the answer	Write the answer
			$10 + $10 + $10 + $10 + $10 = $50 OR 5 × $10 = $50
			$50 + $50 = $100 OR 2 × $50 = $100
			$5 + $5 + $5 + $5 = $20 OR 4 × $5 = $20

Independent practice

1 a–d Teacher to check. Teacher: Look for answers that show students' ability to circle coins that correctly make the designated total and demonstrate that they have a strong grasp of counting with money.

2 a–d Teacher to check. Teacher: Look for answers that show students' ability to circle notes that correctly make the designated total and demonstrate that they have a strong grasp of counting with money.

3 a $2.70 or two dollars and seventy cents

    b $105 or one hundred and five dollars

    c $21.65 or twenty-one dollars and sixty-five cents

    d $35 or thirty-five dollars

4 a $55   b $20

    c $30.45   d $23.40

Order from smallest to largest: b, d, c, a

Extended practice

1 Teacher to check. Teacher: Look for answers that show students' ability to accurately make the given total each time and to use different combinations of numbers.

2 a Possible answers are:
20c; 10c and 10c; 10c, 5c and 5c; 5c, 5c, 5c and 5c

    b Possible answers are:
20c and 5c; 5c, 5c, 5c, 5c and 5c; 10c, 10c and 5c; 10c, 5c, 5c and 5c

UNIT 3: Topic 2

Guided practice

1 50c   **2** 30c   **3** 35c

4 70c   **5** $1.50   **6** 70c

Independent practice

1 Students may draw, write or use equations to show their answers.

    a 8 × 5c coins

    b 4 × 10c coins

    c 2 × 20c coins

2 Students may draw, write or use equations to show their answers.

    a 10 × $10 notes

    b 5 × $20 notes

    c 2 × $50 notes

3 Teacher to check. Teacher: Look for answers that show students' ability to group coins of the same denomination and use skip counting to find the total, or to group coins in easier-to-count groupings, such as $1.

    a 50c   b $1.20   c $12

Extended practice

1 a $2, $1 and 50c
Number of coins: 3

    b 3 × $2, 50c, 20c, 10c and 5c
Number of coins: 7

2 a $55

    b i $35   ii $10   iii $23

3 a $75.95

    b i $55.95   ii $30.95   iii $43.95

UNIT 4: Topic 1

Guided practice

1 a 0, 2, 4, 6, 8 OR 2, 4, 6, 8, 0

    b 0, 3, 6, 9, 2, 5, 8, 1, 4, 7 OR 3, 6, 9, 2, 5, 8, 1, 4, 7, 0

    c 0

Independent practice

1 a and c

4	8	12	16	20	24	28	32	36	40

    b 4

2 a and c

80	75	70	65	60	55	50	45	40	35

    b 5

3 a

10	20	30	40	50	60	70	80	90	100

    b

50	48	46	44	42	40	38	36	34	32

    c

4	9	14	19	24	29	34	39	44	49

    d

30	27	24	21	18	15	12	9	6	3

4 Teacher to check. Look for students who have followed the numbers in the correct sequence.

5 a, b and c

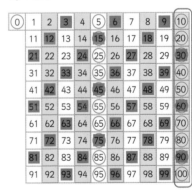

6   10s

7   10, 20, 30, 40, 50, 60, 70, 80, 90, 100

8   30, 60, 90

Extended practice

1   a and c

1	2	③	4	5	6	7	⑧	9	10
11	12	⑬	14	15	16	17	⑱	19	20
21	22	㉓	24	25	26	27	㉘	29	30
31	32	㉝	34	35	36	37	㊳	39	40
41	42	㊸	44	45	46	47	㊽	49	50
51	52	㊼	54	55	56	57	㊽	59	60
61	62	⑥③	64	65	66	67	⑥⑧	69	70
71	72	⑦③	74	75	76	77	⑦⑧	79	80
81	82	⑧③	84	85	86	87	⑧⑧	89	90
91	92	⑨③	94	95	96	97	⑨⑧	99	100

   b   3, 8, 3, 8

   d   2, 5, 8, 1, 4, 7, 0 3, 6, 9

2   a   5s   b   2s   c   10s

3   a   7, 17, 27, 37, 47, 57, 67 OR
       67, 57, 47, 37, 27, 17, 7

   b   10s

UNIT 4: Topic 2

Guided practice

1   a   3 + 4 = 7   b   8 − 3 = 5

   c   9 + 5 = 14

Independent practice

1   a–d   Teacher to check. Teacher:
   Look for answers that show students'
   ability to accurately depict the number
   sentence in a drawing, using the correct
   number of items and identifying the
   operation required.

   Number sentences

   a   15 − 6 = 9

   b   9 − 4 = 5 OR 4 + 5 = 9

   c   13 − 7 = 6

   d   10 + 6 = 16

2   Teacher to check. Teacher: Look for
   answers that show students' ability to
   correctly identify the operation required
   and to think of situations that logically
   demonstrate the operations. Also check
   for appropriate language to match
   addition and subtraction.

3   a   addition   b   addition

   c   subtraction

Extended practice

1   a   Teacher to check. Teacher: Look for
       answers that show students' ability to
       interpret the picture as subtraction
       and write an appropriate story.
       10 − 2 = 8

   b   Teacher to check. Teacher: Look for
       answers that show students' ability
       to accurately interpret the picture
       mathematically – for example, by
       adding the girls and the boys or the
       students with and without hats – and
       to choose the correct operation based
       on their interpretation.

   c   Teacher to check. Teacher: Look for
       answers that show students' ability
       to interpret the picture as addition
       and to write an appropriate story
       with three addends.
       4 + 6 + 7 = 17

UNIT 5: Topic 1

Guided practice

1   a   Teacher to check: approx. 2 hand
       spans

   b   Teacher to check: approx. 8 hand
       spans

   c   Teacher to check: approx. 12 hand
       spans

2   a   12 sticky notes

   b   Students' own answer – approx.
       120 sticky notes

   c   Teacher to check. Teacher: Look
       for answers that show students'
       ability to accurately measure the
       area of their chosen item without
       leaving spaces or overlapping the
       sticky notes.

Independent practice

1   Teacher to check. Teacher: Look for
   answers that show students' ability to
   choose appropriate uniform units of
   length. Also check that students are
   matching one end of their measurement
   unit with the next without any gaps to
   ensure accurate measurement and that
   they line up their measuring tool with
   the edge of the item being measured.

2   Teacher to check. Teacher: Look for
   answers that show students' ability to
   choose an appropriate smaller uniform
   unit to measure the length of the items
   and to accurately measure using their
   chosen unit.

3   match, car, eraser, pencil

4   Teacher to check. Teacher: Look for
   answers that show students' ability to
   choose appropriate uniform units of
   area that will completely cover surfaces
   without gaps. Also check that students
   are not overlapping the units when they
   are measuring area.

5   a   9 squares   b   6 squares

   c   7 squares   d   10 squares

   e   7 squares

6   Figure with area of 10 squares is the
   largest and should be circled.

Extended practice

1   Teacher to check. Teacher: Look for
   answers that show students' ability
   to understand the difference between
   length and area and to choose
   appropriate units to measure both.
   Also check the accuracy of students'
   measurements.

2   The vertical rectangle should be circled.

3   The vertical rectangle should have a
   tick on it.

4   The third rectangle should have a star
   on it.

UNIT 5: Topic 2

Guided practice

1   a–d   Teacher to check. Teacher: Look
   for answers that show students' ability
   to correctly use a ruler starting at 0
   and to record reasonable measurements
   in metres for the given items.

Independent practice

1   Teacher to check. Teacher: Look for
   answers that show students' ability to
   make a reasonable estimate of lengths
   in comparison to a metre, and to then
   accurately measure their chosen items
   to check their answers.

2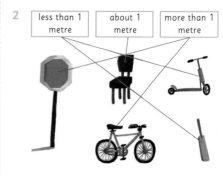

OXFORD UNIVERSITY PRESS

Guided practice

1 a–d Teacher to check. Teacher: Look for answers that show students' ability to correctly use a ruler starting at 0 and to record reasonable measurements in centimetres for the given items.

Independent practice

1 Teacher to check. Teacher: Look for answers that show students' ability to make a reasonable estimate of lengths in comparison to 30 centimetres, and then accurately measure their chosen items to check their answers.

2

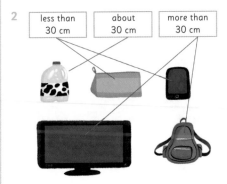

Extended practice

1 a m   b cm   c m
   d m   e cm   f cm
2 swimming pool
3 glass
4 about 2 metres
5 about 20 centimetres

UNIT 5: Topic 3

Guided practice

1 a 6 blocks   b 8 blocks
2 a more than 2 coffee cups
   b less than 2 coffee cups

Independent practice

1 a B, E, A, C, D
   b A, C and D
   c A, B and E
2 a C   b E
   c B and C   d A, D and E
3 a First and third objects should be circled.
   b Second and third objects should be circled.
4 a 1, 2, 3   b 3, 1, 2

Extended practice

1 a and b Teacher to check. Teacher: Look for answers that show students' ability to make reasonable estimates of capacity and to accurately measure and record the capacity of their chosen containers.

2 a and b Teacher to check. Teacher: Look for answers that show students' ability to correctly model an object with a volume of 8 blocks. Accuracy of drawing is difficult, so ensure students are able to explain their drawings to you.

3 a volume   b capacity

UNIT 5: Topic 4

Guided practice

1 The following objects should be circled:
   a birthday cake   b shoes
   c bottle

2 The following objects should be circled:
   a mouse   b car   c pencil

Independent practice

1 B, C, A, E, D

2 a

  b

  c

  d

3 a–c Teacher to check. Teacher: Look for answers that show students' ability to accurately estimate the relative mass of the items in their pairs and use a balance scale correctly to check their answers.

4 a–c Teacher to check. Teacher: Look for answers that show students' ability to make reasonable guesses to identify items with similar, greater and lesser masses than their counters, and to use informal uniform units accurately to check their answers.

Extended practice

1 a the book
   b the shoes and the football
   c the apple
   d the book

2 a–d Teacher to check. Teacher: Look for answers that show students' ability to accurately use uniform informal units to find the mass of each item, and to correctly use a balance scale.

UNIT 5: Topic 5

Guided practice

1 a 12   b 9   c 3   d 6

2 a 9 o'clock   b half past 4

   c quarter past 6   d quarter to 11

Independent practice

1 a   b

  c   d

  e   f

2

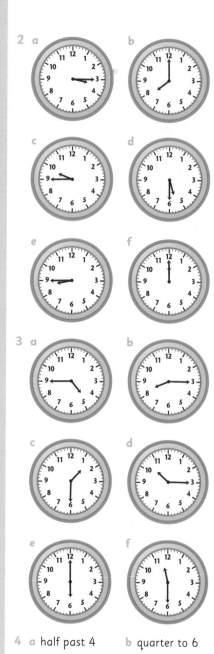

a **b**

c **d**

e **f**

3

a **b**

c **d**

e **f**

4 **a** half past 4   **b** quarter to 6
  c 11 o'clock   **d** quarter past 9
  e half past 12   **f** quarter past 6

Extended practice

1 **a** half past 8   **b** 8 o'clock
  c quarter to 8   **d** half past 9

2 **a** **b**

2:15   3:00
quarter past 2   3 o'clock
OR two fifteen

---

c **d**

3:45   12:00
quarter to 4   12 o'clock

e **f**

7:15   10:00
quarter past 7   10 o'clock
or seven fifteen

UNIT 5: Topic 6

Guided practice

1 **a** Gina
  b Alex
  c Sam

2 **a** 2 minutes
  b 3 minutes

3 **a** 120 minutes
  b 30 minutes

Independent practice

1 **a** 48 hours
  b 14 hours

2 **a** 2 weeks
  b 28 days
  c 3 weeks
  d 70 days

3 **a** 8 weeks
  b 4 weeks and 3 days

4 **a** 4 years
  b 36 months
  c 2 years

5 Students could use this as a group activity and be asked to justify other responses. The most likely answers are below.
  a Seconds: Writing your name
  b Minutes: Eating a sandwich
  c Hours: Sleeping at night
  d Days: Reading a chapter book

---

  e Weeks: The school summer holidays
  f Months: A football season
  g Years: Becoming a top athlete

6 Practical activity. Teacher to check. Look for students who choose an activity that takes the appropriate amount of time.

Extended practice

1 **a** 10 seconds
  b 20 seconds
  c 30 seconds
  d 50 seconds

2 Practical activity. Students could share their strategies to get better at estimating in seconds.

3 Practical activity.

UNIT 5: Topic 7

Guided practice

1 **a** 12
  b 4
  c 3

Independent practice

1

Month	Seasons in the southern hemisphere	Seasons in the northern hemisphere
January	summer	winter
February	summer	winter
March	autumn	spring
April	autumn	spring
May	autumn	spring
June	winter	summer
July	winter	summer
August	winter	summer
September	spring	autumn
October	spring	autumn
November	spring	autumn
December	summer	winter

2 **a–c** Teacher to check. Teacher: Look for answers that show students' ability to accurately sequence months and to match the months to the correct seasons.

OXFORD UNIVERSITY PRESS

Extended practice

1 a autumn

  b late summer and early spring

  c 3

  d

Aboriginal season	Summer, winter, autumn or spring?
High summer	spring, summer
Late summer	summer, autumn
Early winter	autumn
Deep winter	autumn, winter
Early spring	winter
True spring	spring

UNIT 5: Topic 8

Guided practice

1 a Tuesday   b 6th February

  c Monday

2 a 4   b 5

  c Wednesday   d Saturday 30th

Independent practice

1 a

Month	Days
January	31
February	28
March	31
April	30
May	31
June	30
July	31
August	31
September	30
October	31
November	30
December	31

  b April, July, September and December

  c February, March, November

2 a Saturday 18th May

  b Wednesday 22nd May

  c Wednesday, Thursday and Friday

  d Tuesday 14th May

---

e 6

f June

g April

Extended practice

1 a–d Teacher to check. Teacher: Look for answers that show students' ability to correctly identify and write the current month and to accurately label the dates. Also check that students can use the information they have provided to correctly identify the first day of the month and the number of days in the month.

2 a No

  b April, June, September or November

  c 3

  d 17th

UNIT 6: Topic 1

Guided practice

1 a 6 corners   b 6 sides

2 a 5 corners   b 5 sides

3 a 8 corners   b 8 sides

Independent practice

1 a and b

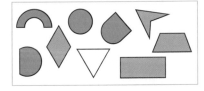

2

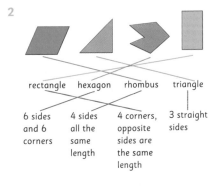

3 Teacher to check. Teacher: Look for answers that show students' ability to use the descriptions to accurately draw a shape that matches the criteria.

Extended practice

1 Note that many shapes have a number of possible classifications.

  a kite, quadrilateral

  b square, quadrilateral, rhombus, parallelogram, rectangle

  c pentagon

  d parallelogram, quadrilateral

  e circle

  f octagon

2

4 corners?	5 corners?	8 corners?	No corners?
kite (or alternative name)	pentagon	octagon	circle
square (or alternative name)			
parallel-ogram (or alternative name)			

UNIT 6: Topic 2

Guided practice

1 a 6 faces   b 12 edges

  c 8 corners

2 a 3 faces   b 2 edges

  c 0 corners

3 a 4 faces   b 6 edges

  c 4 corners

Independent practice

1 a 2 triangles   3 rectangles

  b 2 circles   1 rectangle

  c 1 square   4 triangles

  d 6 squares   0 circles

2 a A, E   b E

  c B, C, D   d C, D   e C

3

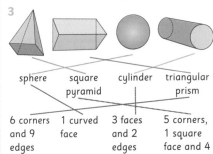

4

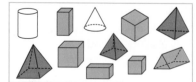

Extended practice

1 and 2

a

cone

b

cube

c

triangular
pyramid

d

triangular
prism

3 **a** rectangular prism

b cylinder

Teacher to check drawings. Teacher:
Look for answers that show students'
ability to accurately represent the given
shapes with the correct shapes in the
faces that are visible.

UNIT 7: Topic 1

Guided practice

1 **a** the tree   **b** the dog

c below the tree OR on the picnic
blanket, or similar

d the bin   **e** the cat

Independent practice

1 **a–d** Teacher to check. Teacher: Look
for answers that show students have
a strong grasp of the vocabulary of
location and are able to accurately
identify where each item is in relation
to other items in the room.

2 Item placement is approximately as
follows:

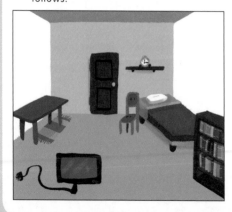

3 **a–d** Teacher to check. Teacher:
Look for answers that show students'
ability to accurately use terms such as
"next to", "to the left of", "between"
and "opposite".

4 **a** Either the toy shop, book shop,
toilets and play area OR the food
court, muffin shop, jewellery store
and hairdresser depending on
route chosen.

b left

Extended practice

1 **a** the emu   **b** the koala

2 Teacher to check. Teacher: Look for
answers that show students' ability to
use locational language to accurately
describe the route and directions that
can be followed.

UNIT 7: Topic 2

Guided practice

1 **a** slide   **b** flip   **c** flip   **d** slide

Independent practice

1 **a** 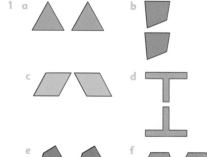 **b**

c   **d**

e   **f**

2 **a** slide   **b** either   **c** flip
d slide   **e** either   **f** flip

Extended practice

1 **a**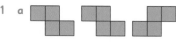

b

c

d   **e**

UNIT 7: Topic 3

Guided practice

1 **a** half turn   **b** quarter turn
c quarter turn   **d** quarter turn
e half turn   **f** quarter turn

Independent practice

1 **a** quarter turn

b half turn

c half turn

d quarter turn

2 **a** **b**

c **d**

3 **a** half turn   quarter turn

 OR

b half turn   quarter turn

 OR

Extended practice

1 **a** quarter turn

b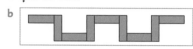

2 **a** half turn, then quarter turn to
the right

b

3 **a** The third and fifth shapes should
be circled.

b The first and second shapes
should be circled.

OXFORD UNIVERSITY PRESS

UNIT 8: Topic 1

Guided practice

1

Ice-cream flavours	Number of students (ticks)	Total (number)
Vanilla	✓✓✓✓✓✓✓✓ ✓✓✓✓✓✓✓✓	16
Chocolate	✓✓✓✓✓✓✓✓ ✓✓✓✓✓✓✓ ✓✓✓	18
Mint	✓✓✓✓✓✓✓✓ ✓✓✓✓	12
Strawberry	✓✓✓✓✓✓✓✓	8

2 Teacher to check. Teacher: Look for answers that show students' ability to ask an open-ended question that will get the responses listed in the table, for example, "What is your favourite ice-cream flavour?"

Independent practice

1

Do you have a brother?	
Yes	No
10	8

2a

What pet do you have?	
Cat	6
Dog	10
Reptile	2
Other	5
None	3

b 26

3 Teacher to check. Teacher: Look for answers that show students' ability to accurately record 12 answers using either the tick method or by writing down students' names.

4 Teacher to check. Teacher: Look for answers that show students' ability to accurately record 12 answers using an appropriate method.

5 Teacher to check. Look for answers that show students' ability to ask an open-ended question that will get the responses listed in the table – for example, "What is your favourite sport?"

6 a–b Teacher to check. Answers will vary depending on student data. Teacher: Look for answers that show students are able to interpret their data accurately to find the most and least popular sport.

Extended practice

1 Teacher to check. Teacher: Look for answers that show students' ability to choose an appropriate question that can only have "yes" or "no" as the answer.

2 Teacher to check. Teacher: Look for accurate recording of both the question and the results in the table.

3 Teacher to check. Teacher: Look for recording strategies such as ticks or tally marks. Ensure the results match the results that students recorded in question 2.

UNIT 8: Topic 2

Guided practice

1

	Milk	Water	Orange juice	Soft drink
Total	3	8	15	17

2

	Red	Blue	Green
	ⱵⱵⱵ ⱵⱵⱵ ‖	ⱵⱵⱵ ⱵⱵⱵ ‖	ⱵⱵⱵ ‖‖‖
Total	12	11	8

Independent practice

1 a Teacher to check. Teacher: Look for answers that show students' ability to choose appropriate categories, such as shape or colour, and who can identify variables that match – for example, circles and rectangles for shape and blue and green for colour.

b Teacher to check. Teacher: Look for answers that show students' ability to make an accurate tally and to use tally mark groupings correctly.

c Teacher to check. Teacher: Look for answers that show students' ability to accurately count their tally marks.

d Teacher to check. Teacher: Look for students who are able to draw simple conclusions from their data.

2 a Teacher to check. Teacher: Look for answers that show students' ability to choose appropriate variables that are likely to appeal to the classmates being surveyed – for example, basketball, netball, football, cricket – and who can record the variables in the correct section of the table.

b Teacher to check. Teacher: Look for answers that show students' ability to use tally marks correctly to keep track of responses.

c Teacher to check. Teacher: Look for answers showing totals that match the tally marks they recorded.

3 a and b Teacher to check. Teacher: Look for answers that show students' ability to correctly identify the most and least popular options using the data they collected.

c Teacher to check. Teacher: Look for answers that show students' ability to come up with plausible options that their classmates are likely to choose, such as football or rugby.

Extended practice

1 a Teacher to check. Teacher: Look for answers that show students' ability to identify variables that match the pictures – for example, number of legs, animals that can and cannot fly, colours of animals or animals that live in the water/on land.

b Teacher to check. Teacher: Look for answers that show students' ability to construct a table with the correct number of columns and rows to record their variables and results. Also check that students are able to make an accurate tally that matches the data based on their chosen categories.

c Teacher to check. Teacher: Look for answers that show students' ability to accurately count and record the totals of their tally marks and to write the total in the correct place in their table.

UNIT 8: Topic 3

Guided practice

1 a ants b worms c 4 d 24

Independent practice

1 a

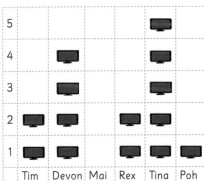

5						
4		▬			▬	
3		▬			▬	
2	▬	▬		▬	▬	
1	▬	▬		▬	▬	▬
	Tim	Devon	Mai	Rex	Tina	Poh

b Tim and Rex c Tina
d Mai

2 a

Hair colour	Tally	Total
Brown	IIII I	6
Blond	IIII	4
Black	IIII III	8
Red	II	2

b

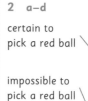

8			😀	
7			😀	
6	😀		😀	
5	😀		😀	
4	😀	😀	😀	
3	😀	😀	😀	
2	😀	😀	😀	😀
1	😀	😀	😀	😀
	Brown	Blond	Black	Red

3 a Teacher to check. Teacher: Look for answers that show students' ability to understand how to record data in a list and that have 10 pieces of data recorded.

b Teacher to check. Teacher: Look for answers that show students' ability to correctly label the table, whose tally marks match the data in their list and whose total matches the tally marks.

c Teacher to check. Teacher: Look for answers that show students' ability to match the representations in their pictographs to those in their tally tables.

d Teacher to check. Teacher: Look for answers that show students' ability to make a statement that accurately matches the data, such as identifying the category with the greatest or least number of responses, or comparing the numbers in both categories.

Extended practice

1 a

Transport	Tally	Total
Car	IIII IIII	9
Bike	IIII	5
Walk	IIII II	7
Bus	I	1

b 5 **c** walk **d** 2

e Teacher to check. Teacher: Look for answers that show students' ability to write a question that directly relates to the data, such as how many people use a particular form of transport.

UNIT 9: Topic 1

Guided practice

1 a and b Teacher to check. Teacher: Look for answers that show students' ability to justify their selections and use the language of chance to describe the probability of each event.

2 a certain (unless there are special circumstances where students do not wear shoes all day)

b certain

Independent practice

1 a, b, d and e Teacher to check. Teacher: Look for answers that show students' ability to justify their selections using the language of chance.

c certain   **f** impossible

2 a most likely (for most students)

b certain   **c** less likely

d impossible

3 a–d Teacher to check. Teacher: Look for answers that show students' ability to choose and describe events that accurately match each chance term, and to demonstrate that they understand the nuances between related terms, such as "certain" and "most likely".

Extended practice

1 a most likely   **b** less likely

   c impossible

2 a–d

certain to pick a red ball

impossible to pick a red ball

less likely to pick a green ball

most likely to pick a blue ball

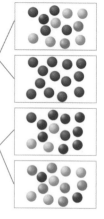

OXFORD UNIVERSITY PRESS